Ingo Rotscholl

Spectral near field data of LED systems for optical simulations

SPEKTRUM DER LICHTTECHNIK BAND 17

Lichttechnisches Institut
Karlsruher Institut für Technologie (KIT)

Spectral near field data of LED systems for optical simulations

by
Ingo Rotscholl

KIT Scientific Publishing

Karlsruher Institut für Technologie
Lichttechnisches Institut

Spectral near field data of LED systems for optical simulations

Zur Erlangung des akademischen Grades eines Doktor-Ingenieurs von der KIT-Fakultät für Elektrotechnik und Informationstechnik des Karlsruher Instituts für Technologie (KIT) genehmigte Dissertation

von M.Sc. Ingo Rotscholl geboren in Lutherstadt Wittenberg

Tag der mündlichen Prüfung: 7. Dezember 2017
Referent: Prof. Dr. Cornelius Neumann
Korreferenten: Prof. Dr. Wilhelm Stork, Prof. Dr. Franz Schmidt

Impressum

KIT Scientific Publishing

Karlsruher Institut für Technologie (KIT)
KIT Scientific Publishing
Straße am Forum 2
D-76131 Karlsruhe

www.ksp.kit.edu

Print on Demand 2018 – Gedruckt auf FSC-zertifiziertem Papier

ISSN 2195-1152
ISBN 978-3-7315-0750-5
DOI 10.5445/KSP/1000078518

Kurzfassung

Diese Arbeit stellt ein allgemeines Mess- und Modellierungsverfahren zur Erzeugung orts- und winkelaufgelöster Spektralverteilungsmodelle für LEDs vor, wie sie im Entwurfsprozess moderner und hochwertiger LED-basierter Leuchten zunehmend benötigt werden. Im Gegensatz zu bestehenden Methoden ist das hier vorgestellte Verfahren dabei nicht auf zeitaufwendige winkelaufgelöste Spektralmessungen angewiesen. Die Anzahl der nahfeldgoniophotometrischen Messungen wird durch das Messobjekt bestimmt und stets minimiert.

Das Grundkonzept der „Physikalisch motivierten Basisspektren" (PMBS) basiert auf der Annahme, dass sich jedes vom Ort oder Winkel abhängige Spektrum als gewichtete Summe der im System vorkommenden Basisspektren beschreiben lässt. Unter Hinzunahme der spektralen Information des goniophotometrischen Messsystems ergibt sich das spektrale Nahfeldmodell als Lösung eines simplen linearen Gleichungssystems der nahfeldgoniophotometrischen Messungen. Die Zusammenhänge können direkt auf Fernfeldmessungen bzw. andere kamerabasierte Messungen übertragen werden.

Die Überführung von PMBS in die Praxis besteht aus drei Teilen, welche lediglich leicht verfügbare physikalische Informationen, wie die Anzahl unterschiedlicher spektraler Quellen, die räumliche Trennung verschiedenfarbiger LED-Chips oder die physikalische Unmöglichkeit negativer Spektren, verwenden. Das Verfahren wird anhand verschiedener aktueller LED-Systeme, wie einer leuchtstoffbasierten weißen LED, einer RGB LED einer rot/weißen LED und einer RGBW LED validiert. Die erhaltenen Modelle können direkt in modernen kommerziell verfügbaren Strahlverfolgungsprogrammen verwendet werden.

ABSTRACT

This thesis presents and validates a fast, accurate and general measurement and modeling technique to obtain spectral near field data of LED systems in order to improve the optical design process of modern high quality LED systems. It requires only a minimum of goniophotometric near field measurements as well as no time-consuming angularly resolved spectral measurements. The required measurement effort is thus much smaller than in existing state-of-the-art methods and the approach works for a wide range of modern LED systems.

The procedure is named physically motivated basis spectra (PMBS) as its main assumption is that each piece of angularly and spatially varying spectral information can be described as the weighted sum of its physical basis spectra such as the individual semiconductors or a phosphor. Based on detailed spectral information regarding the goniophotometric measurement setup, the spectral model is obtained by solving a simple system of linear equations using the obtained near field measurements. PMBS can also be applied on far field measurements or camera based measurements.

The proposed workflow to create spectral near field data consists of three major steps, which utilize physically available information such as the number of different spectral sources, the spatial separation of colored LEDs from each other or the physical impossibility of negative spectra, to overcome their practical challenges. The complete process is validated and applied to different state-of-the-art LED systems, such as a phosphor converted white LED, a RGB LED, a Red/White LED and a RGBW LED system. The obtained results can be used directly in state-of-the-art ray tracers.

Contents

CHAPTER 1

INTRODUCTION

The sole requirement of the first commercially used artificial light sources - gas lights - was to provide light. However, starting with the development of incandescent lamps, artificial light sources began to compete in economic aspects such as production costs, lifetime and energy efficiency. Both technologies improved and soon artificial light sources became common and the field of general lighting started to grow. During the 1960s and 1970s, in the shadows of the established light sources, another light source technology was invented [1]. This was the starting point of the development of light emitting diodes (LEDs) [2], [3].

The basis of the LED breakthrough originated in 1994. It was the invention of the first high brightness blue LED [4] by Nakamura et. al. as it enabled white LED based lighting. The LED technology rapidly improved in the following years and quickly started to be competitive in terms of lifetime and energy efficiency. Nowadays, LEDs have become one of the most important light sources in lighting technology. LED lighting applications range from medical [5], [6], horticulture [7] and automotive lighting [8] over several general lighting applications such as street lighting, work place lighting as well as shop and home lighting [9]. It is expected that the importance of LEDs will further increase in the next few years [3], [10].

However, not only the direct efficiency of the light source itself helped LEDs to become such a success story. Their small sizes and comparable low operation temperatures also enabled the possibility of placing cost-efficient plastic optics near to the LED source [11]. Examples are light guides, free-form lenses or metal coated free-form reflectors. A feasible optical design of these free-form elements ensures a high optical system efficiency, specific photometric quantities and a compact system size. However, the design of these compact systems requires the optical designer to include the near field radiation properties of each light source in the system [12], [13]. This means the light source model has to describe the direction of the light rays coming from the source and their starting points on the spatially extended source. This information is essential for designing specific, efficient and robust high quality optical systems.

The measurement technique which provides this near field information is called near field goniophotometry [14], [15]. The measurement result obtained is called ray file. It can be used in commercial ray tracing software [16], [17]. This software enables to trace light rays according to the rules of geometrical optics and thus to assess an optical design. Therefore, near field goniophotometry, ray files and ray tracing have become important optical measurement and simulation tools which enable the design of efficient high quality lighting systems considering the near field of LEDs [12], [18].

However, in modern lighting technology, "lighting quality" refers to a permanently increasing number of different aspects, which also depends on the specific application. Besides photometric quantities and efficiency this may include visual and aesthetic properties such as uniformity, color uniformity and color rendering or non-visual effects of light on human beings called human centric lighting [2], [3], [19]. Examples are the influence of the color temperature on the day and night rhythm [20], [21], the concentration during work or the learning process of school children [22].

All these additional quality aspects are directly related to the spectrum. Different material systems and LED chip designs provide different spectral power distributions. These can be combined with different phosphors, which convert spectral parts of the light from the LED chip into other spectral parts. This enables the design of specific spectral power distributions. These distributions can be optimized in terms of several different quality aspects such as efficiency, chromaticity, color rendering or other color appearance metrics as well as non-visual aspects [23]–[26]. At the same time, the desired uniformity requirements have led to the need for efficient mixing optics for different LEDs to ensure a high lighting quality in all the above mentioned aspects [27]–[29].

However, the optical design of color mixing optics as well as the simulation of complete spectral distributions in an optical system within a ray tracer is a challenging task. The simulations are not as precise as photometric simulations and require a comparably large amount of computation time [30]. Furthermore, in contrast to conventional incandescent or fluorescent lamps, the required modern LED multi-chip systems as well as plain white LEDs can show strong spectral variations in both the angular and the spatial dimension. These variations can also differ for individual LEDs due to manufacturing tolerances [16], [31]. Thus, the optical design requires a accurate LED models which precisely describe the complete spectral radiation in each direction from each point of the physical light sources. These models can be referred to as hyperspectral LED models.

Since conventional near field goniophotometry does not directly provide spectrally resolved near field information, there is a lack of suitable hyperspectral LED light source models for the ray tracing simulations. Nevertheless, if the optical design is based on wrong assumptions regarding the light source properties, the best design and the most precise and time-consuming simulation will still lead to an unsatisfactory result. This leads to a need for precise hyperspectral LED models in terms of the angular and the spatial dimension.

1.1 AIMS OF THE THESIS

The main aim of this thesis is to propose, develop and validate a measurement and modeling technique to obtain hyperspectral LED models which can be used in conventional non-sequential ray tracing software. In contrast to the existing state of the art, this method has to combine a suitable measurement time with acceptable spectral deviations in the context of lighting technology. This also includes the required measurement systems, as well as the model assumptions and the deduced limitations so that the method can be applied to state-of-the-art LED systems. The models have to be capable of simulating a spectrally varying distribution, which may be based on an angularly and spatially varying LED source and a spectrally sensitive optical system. Examples of such distributions are shown in the photographic images of Fig. 1.1.

(a) **(b)**

Figure 1.1: Examples of spectrally varying LED based lighting:
(a) Plain phosphor converted white LED.
(b) Optical system consisting of an spectral-angularly and spectral-spatially varying source and a dispersion prism.

1.2 STRUCTURE OF THE THESIS

The thesis develops, applies and validates one specific method to obtain hyperspectral LED models. The fundamentals, which are required in the thesis, are briefly summarized in chapter 2 which mainly focuses on state-of-the-art near field goniophotometry and some aspects of ray tracing and optical design as well as light source models. Furthermore, spectral properties of LEDs as well as selected colorimetric terms and concepts are briefly explained.

Chapter 3 introduces first the concept and application of spectral ray files. Afterwards, the existing state of the art is summarized. The concept of each method is qualitatively explained and assessed with respect to earlier deduced measurement properties.

Chapter 4 serves as the theoretical basis for the following chapters. It motivates, introduces and mathematically describes the physically motivated basis spectra (PMBS) approach, which extends selected concepts from the state of the art of the previous chapter. Based on the mathematical description of PMBS, a theoretical modeling workflow as well as fundamental limitations are concluded. Finally, the main practical challenges are summarized at the end of the chapter.

Chapters 5- 7 each focus on one of the individual challenges of PMBS, concerning its practical application. Thus, each main challenge is considered individually in theory and the deduced solutions are explained and supported with practical examples. While chapter 5 concentrates on obtaining spectral LED model parameters, chapter 6 considers typical uncertainties of the required measurements to adapt the measurements and optimize the data base. Finally, chapter 7 describes the construction of the hyperspectral LED models with the obtained measurement data in detail.

A validation of the complete workflow as well as a comparison with an important state-of-the-art method is provided in chapter 8. This also includes the spectral ray tracing of sources, whose spectrum changes as a function of angular and spatial dimension, combined with a spectrally sensitive system.

Chapter 9 concludes and discusses the developed methods, problems, limitations and obtained results with respect to the desired application and in a broader context. The main findings of the thesis are briefly summarized in the last chapter 10. Furthermore, it suggests concepts which may improve both PMBS and near field goniophotometry in the future.

CHAPTER 2

FUNDAMENTALS

The aim of this chapter is to summarize the necessary fundamentals regarding the spectral characteristics of LEDs, photometry and colorimetry. Furthermore, ray tracing and its typical light sources as well as near field goniophotometry are briefly introduced. A more complete and general overview of the fundamentals is given in the cited literature.

2.1 SPECTRAL PROPERTIES OF LEDS

A Light Emitting Diode (LED) is an optoelectronic semiconductor converting electrical energy into incoherent radiation. A part of the electrical energy is also converted into thermal energy. The emitted radiant power distribution in watt per frequency or wavelength λ is called spectral power distribution $S(\lambda)$ or in short spectrum. The spectrum depends on the specific material system of the LED as well as its thermal and electrical operation conditions. The wavelength at which the highest radiation occurs is called peak wavelength. In this thesis, the peak wavelength is the only characteristic wavelength associated with an LED spectrum. Examples of LED spectra and their peak wavelengths are given in Fig. 2.1. It is common in lighting technology to visualize all quantities as a function of the wavelength in vacuum rather than the frequency.

The spectral width of LED spectra prohibit the direct generation of white light using solely one semiconductor chip. However, as the generation of white light is an important aspect in lighting technology, two possibilities for creating white LED based light have been developed. One is the additive mixture of different LED spectra with spatially separated LED chips, for instance red, green and blue. The color is tuned by adapting the power of the individual chips. However, the color mixing of the spatially separated chips is a challenging task.

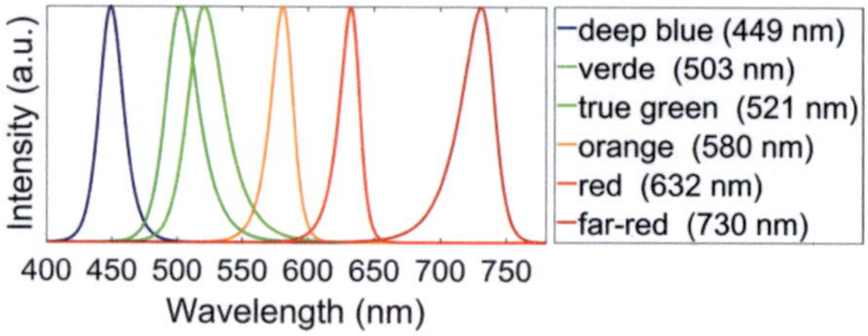

Figure 2.1: Examples of typical LED spectra (spectra and labeling according to [32]–[35]).

White light can also be created by combining typically a blue or ultraviolet LED chip with one or several conversion phosphors. The phosphor partly absorbs the high-energy LED radiation and re-emits radiation with longer wavelengths. Due to the additional scattering properties of the phosphor, the resulting radiation is partly mixed. The thickness of the phosphor layer can be used to tune the light color. A thicker layer results in a yellower or warm white lighting. Figure 2.2 (a) shows white LED spectra, which are based on a RGB LED system, and Fig. 2.2 (b) shows spectra, which are based on phosphor converted LED systems. Nevertheless, depending on the specific design of the phosphor coating, the resulting spectra may have an angular color uniformity. This results from an angular dependent "LED radiation" to "phosphor radiation" ratio caused by deviating phosphor conversion lengths. An example of this phenomenon is discussed in detail in section 8.1. More information on LEDs, LED systems, and conversion phosphors is provided in [36], [37].

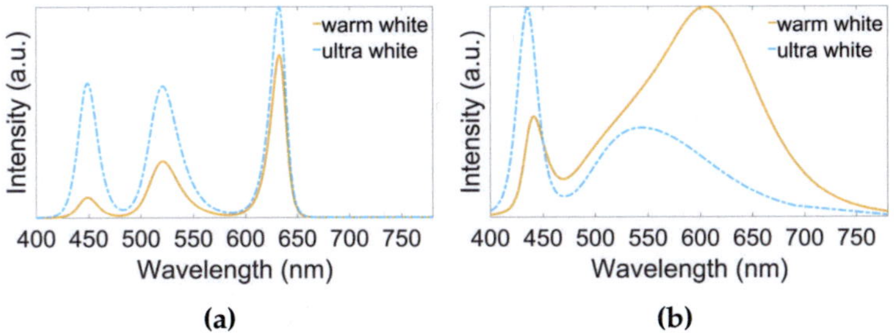

Figure 2.2: White LED spectra based on (a) additive mixture of red, true green and deep blue (according to Fig. 2.1) and (b) phosphor converted LED spectra according to [38], [39].

2.2 RADIOMETRY, PHOTOMETRY AND COLORIMETRY

Radiometry is the science of measuring the optical part of electromagnetic radiation ranging from ultraviolet to infrared [40]. Photometry can be considered an area of radiometry, which focuses solely on the visible part of the spectrum. Another important difference between them is the application of specific actinic action spectra. Actinic action spectra are weighting functions, which are used to weight the measured spectral power distribution according to photobiological or photochemical effects. In radiometry, there is no specific weighting function. In photometry, the $v(\lambda)$ function, which is used to model the sensitivity of the human eye, is the most important weighting function. If an actinic action spectrum or any other spectral weighting function is applied, the spectrally integrated value changes. In terms of the $v(\lambda)$ function, the resulting quantities have specific names such as luminous power, light intensity, illuminance or luminance to distinguish them from their radiometric counterparts such as radiometric power, radiant intensity, irradiance or radiance. In this thesis, all arbitrary weighted and spectrally integrated values obtained with any function different from the $v(\lambda)$ function are marked with the index "pseudo" to distinguish them from the unweighted radiometric and photometric values. More information on the basis quantities of radiometry and photometry can be found in [40], [41].

Two other important actinic action spectra are the color matching functions $\overline{x}(\lambda)$ and $\overline{z}(\lambda)$, which are related to human color perception. Similarly to the $v(\lambda)$ function, they were defined by the Commission Internationale de l'Eclairage (CIE). The color matching functions $\overline{x}(\lambda)$, $\overline{z}(\lambda)$ and the $\overline{y}(\lambda) = v(\lambda)$ function are the basis of the CIE 1931 color space. The integration of a spectrum $S(\lambda)$, which is weighted with a color matching function, results in a tristimulus value X_{CIE}, Y_{CIE}, or Z_{CIE}:

$$\begin{aligned} X_{\text{CIE}} &= \int S(\lambda) \times \overline{x}(\lambda) d\lambda \\ Y_{\text{CIE}} &= \int S(\lambda) \times \overline{y}(\lambda) d\lambda. \\ Z_{\text{CIE}} &= \int S(\lambda) \times \overline{z}(\lambda) d\lambda \end{aligned} \tag{2.1}$$

The tristimulus values serve as an important starting point for the science of color measurement called colorimetry. An important light color representation used in this thesis is the CIE 1976 uniform chromaticity scale diagram or in short CIE $u'v'$ diagram. The coordinates are the so called chromaticity coordinates u' and v', which are calculated according to

$$\begin{aligned} u' &= \frac{4X_{\text{CIE}}}{(X_{\text{CIE}} + 15Y_{\text{CIE}} + 3Z_{\text{CIE}})} \\ v' &= \frac{9Y_{\text{CIE}}}{(X_{\text{CIE}} + 15Y_{\text{CIE}} + 3Z_{\text{CIE}})} \\ \Delta u'v' &= \sqrt{(u'_1 - u'_2)^2 + (v'_1 - v'_2)^2} \end{aligned} \tag{2.2}$$

In contrast to a similar diagram, the CIE 1931 xy diagram, the euclidean distance $\Delta u'v'$ between two points u'_1, v'_1 and u'_2, v'_2 in the CIE $u'v'$ diagram is approximately proportional to the perceived difference of their light color stimuli. The CIE thus recommends evaluating white light color differences [42], [43] and also saturated light color differences with this diagram [43]. The color matching functions and the CIE $u'v'$ diagram are shown in Fig. 2.3.

Not all colorimetric values are based on the tristimulus values. The color rendering index (CRI) and its suggested successors such as TM-30-15 (Technical Memorandum-30-15), which are all used to rate the capability of a light spectrum to render object colors, require the complete spectrum $S(\lambda)$ because they are based on different spectral weighting functions. A comprehensive overview on this topic is provided in [44].

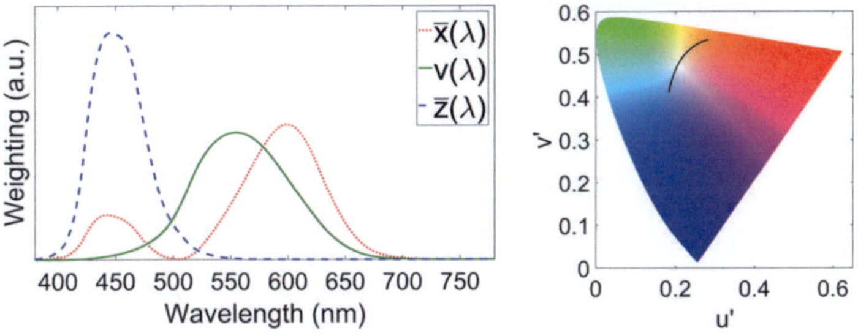

(a) **(b)**

Figure 2.3: Color matching functions (a) and CIE $u'v'$ diagram (b).

An important phenomenon in colorimetry is known as metamerism. It refers to the effect that different spectral power distributions may result in the same color stimulus. The mathematical explanation is that the integration of the weighted spectral power distributions may result in the same relations between the tristimulus values X_{CIE}, Y_{CIE}, and Z_{CIE}. While different chromaticity coordinates cannot have the same underlying spectrum, different spectra may lead to the same chromaticity coordinates. These colors are called metamers. An example are the white RGB based spectra in Fig. 2.2 (a), which are assigned to the same chromaticities as the phosphor converted white LED spectra in Fig. 2.2 (b). More information on colorimetry and more advanced concepts such as color appearance models are provided in [45], [46].

2.3 RAY TRACING

The precise propagation of each light source's electromagnetic radiation can be described with Maxwell's equations. However, solving these differential equations is both computationally expensive and unnecessary in the typical optical design process of lighting technology. In lighting technology, the spatial dimensions of the optically relevant geometries are usually much larger than the wavelengths of the usually incoherent radiation and effects as interference or diffraction can be neglected without affecting the precision [12]. In these cases, the mathematical assumption $\lambda \rightarrow 0$ is valid and leads to geometrical optics [47]. Then the light can be approximated in the form of rays, which are perpendicular to the wave front and only the rules of geometrical optics such as Snell's law of refraction, the law of reflection or the Fresnel equations are considered if a ray encounters a material interface [48].

A ray tracer is a software tool, which facilitates the application of the rules of geometrical optics[1] on up to several billion rays within a typically three dimensional opto-mechanical setup. Besides the geometries, all optical material properties such as refractive indices, dispersion relations or surface parameters are usually part of the calculations. Ray tracers are typically used to design optical elements such as lenses, reflectors or total internal reflection optics in such a way that they provide a specific light distribution or illumination in combination with a specific light source or to predict the visual perception of an

[1] Ray tracers may also support physical optics calculations.

illumination scenario. Therefore, light distributions can be evaluated regarding their radiometric, photometric or colorimetric properties. Furthermore, system analyses can be performed, for instance to visualize specific ray paths called sequences within the opto-mechanical setup. A typical ray tracing setup consists of a source, which serves as ray origin, the opto-mechanical setup and sensors, which evaluate the tracing results. An example is shown in Fig. 2.4.

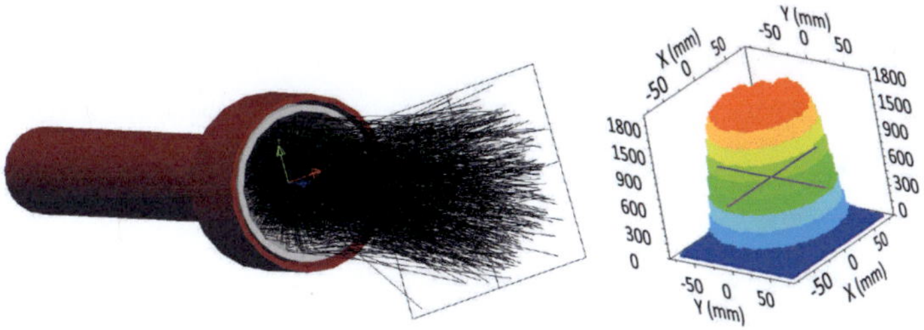

Figure 2.4: Ray tracing setup of a flash lamp and irradiance distribution in the ray tracer LightTools.

There is sequential and non-sequential ray tracing. In sequential ray tracers, the sequence of the optically active geometries is predefined. It is faster and typically used to design imaging optics. In contrast, non-sequential ray tracing has no predefined tracing sequence. It is typically used to design non-imaging optics or to simulate effects such as stray light. There are also cases in which both approaches are combined. An example is the design of modern head-up-displays [49].

In direct non-sequential ray tracing, rays start at a defined light source and are evaluated at defined sensors. In inverse ray tracing, the rays start at the sensor and propagate to the sources. Inverse ray tracing can be used if only a small number of rays from the source reach the sensor. In both cases, the rays are traced through the optical system until they are absorbed or until they have no more interactions with the optical system. In this thesis, the ray tracer OptisWorks is used to evaluate the light source models.

Several ray tracers, for example OptisWorks, use a Monte Carlo approach. This means that each ray is either completely reflected, completely transmitted or completely absorbed with a specific probability, which is defined by the rules of geometrical optics. This also means that if the same ray is traced twice, its sequence throughout the optical system may change. This has the consequence that the number of rays is an important precision criterion of the simulation. The statistical error of a Monte Carlo ray tracer is indirectly proportional to the square root of the number of rays used.

Typically, the optical design process consists of several steps such as the initial design and the design optimization [12]. The initial design provides a first geometry, which is based on simplified assumptions. It is followed by the design optimization, which aims to optimize the initial geometry by considering more realistic conditions. An important aspect in each step of the optical design is to choose an appropriate light source model. There are several types of light source models in ray tracers, which will be introduced below. More information regarding the optical design process and optical design is provided in [12], [48].

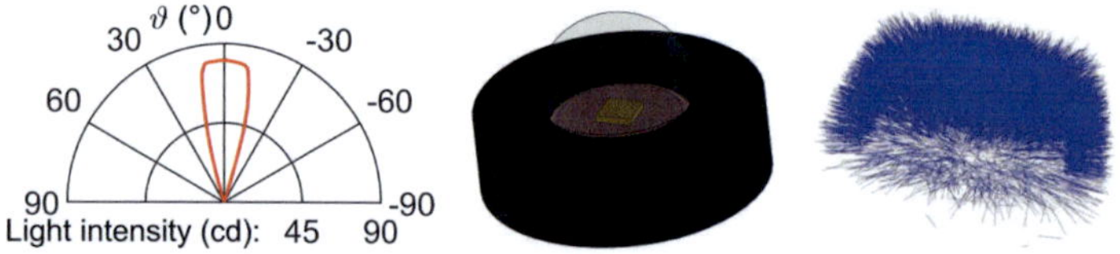

(a) Point source **(b)** Geometrical model **(c)** Ray file

Figure 2.5: Typical light source models used in ray tracers.

The point source model shown in Fig. 2.5 (a) is the most simplified light source model. The extensions of the light source are ignored and it is thus only described by its light intensity distribution curve. The error associated with this approximation is usually neglected, if the distance between the source and the observer or the optical element is beyond the photometric limiting distance. More information on this distance is provided in [50]. Although it is the least precise model, it is often chosen because it enables the use of advanced mathematical optical design techniques such as the supporting quadric method [12], [51] or tailoring [51] during the initial design step [12]. The point source model may be based on a solely mathematical model, a goniophotometric measurement or an indirect camera measurement [52].

Geometrical light source models are a possibility for modeling extended light sources [12]. They extend from simple models, for instance an extended plane, to complex physical CAD models with defined optical parameters of the different materials. A schematic example is provided in Fig. 2.5 (b). Often, the core of these models is another light source model or a mathematical description for instance a Lambertian emitting plane. Radiance or luminance maps may be used as source models as well [53], [54]. They are created by ray tracing of another light source model. These models speed up the ray tracing process and allow both direct and inverse ray tracing.

A ray file as shown in Fig. 2.5 (c) models an extended light source as well. It represents the near field of the source as a list of vectors. Each vector represents a ray with a three dimensional starting point and a normalized direction [16]. The luminous or (pseudo-)radiometric flux is either provided by the density of the rays with constant amplitudes or an individual amplitude for each ray. There are several different ray file formats for the individual ray tracing softwares. Some formats allow the inclusion of additional information like the wavelength or the polarization. In 2013, the TM-25 format was defined by the Illuminating Engineering Society (IES) as a general standard for the ray file format [17], [55]. Ray files are typically used in the optical design process if the extension of the source can no longer be neglected. However, they allow only direct ray tracing. Ray files are typically provided on the home page of the light source manufacturers. They are usually based on near field goniophotometric measurements [14], [15].

2.4 NEAR FIELD GONIOPHOTOMETRY

Ray files are usually based on a set of luminance or pseudo-radiance images, which are measured goniometrically [56]. In this measurement technique, the luminance images are obtained from different angular positions. These measurements are performed with a so called near field goniophotometer[2], as shown in Fig. 2.6 (a). A typical near field goniophotometer consists of at least two moving axles, an imaging luminance measurement device (ILMD[3]) - also referred to as a luminance or color camera -, a photometer and sometimes also a spectroradiometer. In the ILMD, an objective (lens) images the luminance of the device under test on a spatially resolved sensor, for instance a CCD (charge-coupled device) or CMOS (complementary metal-oxide-semiconductor). An additional optical transmission filter adapts the spectral sensitivity of the system to the $v(\lambda)$ function. The ILMD

[2] Strictly speaking, it is more often used as near field gonio(pseudo-)radiometer in this thesis. Nevertheless, in accordance with the commonly used terminology, the terms goniophotometry and goniophotometric are used in these cases as well.

[3] Strictly speaking, it is more often used as imaging (pseudo-)radiometric measurement device in this thesis. Nevertheless, in accordance with the commonly used terminology the abbreviation ILMD is used in these cases as well.

provides the luminance and pseudo-radiance images and thus the near field information. The spectral information is lost due to the optical integration on the sensor.

One or several filter wheels within the ILMD may allow the replacement of the $v(\lambda)$ matching filter by another filter to adapt the spectral sensitivity of the system to other actinic action spectra such as $\overline{x}(\lambda)$ and $\overline{z}(\lambda)$. To reduce the complexity in the manufacturing process, the $\overline{x}(\lambda)$ adaption filter is sometimes represented by two filters. They are named $x_{\text{short}}(\lambda)$ and $x_{\text{long}}(\lambda)$ in this thesis. A schematic ILMD with a filter wheel is visualized in Fig. 2.6 (b). These systems can be used to apply selected actinic action spectra directly on the measured spectra and thus to obtain a ray file set, which represents the selected quantity. However, these ray files are not suited to perform spectral simulations, which will be discussed in more detail in section 3.2.

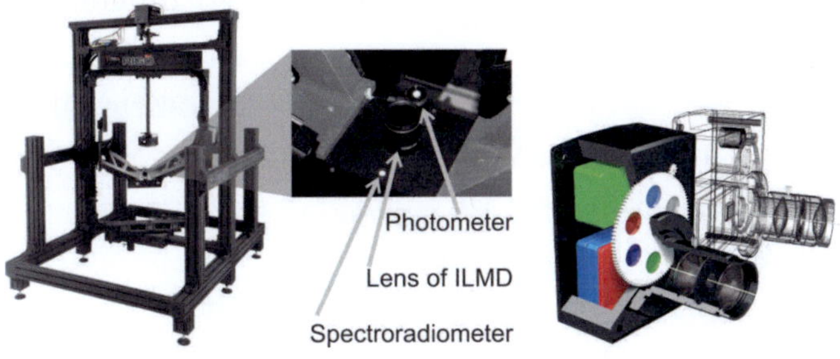

(a) (b)

Figure 2.6: Near field goniophotometer with typical sensors (a) and schematic image of an ILMD with filter wheel (b), (images reprinted with permission from [57], [58]).

The photometer as well as the spectroradiometer provide no spatial resolution. Although they are mounted on a near field goniophotometer, the source models obtained solely with these sensors can be considered as far field models. The photometer's system sensitivity is also adapted to the $v(\lambda)$ function. In near field goniophotometry the photometer, which also provides no spectrally resolved information, is used to obtain the absolute luminous flux of ray files, which are based on a $v(\lambda)$ ILMD-based goniophotometric measurement.

The spectroradiometer is the only sensor which provides spectrally resolved information. There are array and scanning spectroradiometers. Although a scanning spectroradiometer offers a higher spectral resolution and a larger stray light suppression, which leads to a higher dynamic range and therefore a smaller measurement uncertainty [59], its higher measurement time is unfavorable in combination with a goniometer. This is why the faster but less precise array spectroradiometers are typically used for goniometric measurements. The measurement time of an angularly resolved goniometric spectral measurement is still very high compared to a similarly resolved ILMD-based measurement due to the difference of the integration times ultimately leading to a start-stop-motion of the goniometric spectral measurement [59].

In this thesis, the "RIGO801 near field goniophotometer" from the TechnoTeam Bildverarbeitung GmbH was used to perform the near field measurements. It is equipped with a photometer from the Czibula & Grundmann GmbH, a "LMK5 color" camera from TechnoTeam and a "specobs 1211" spectroradiometer from JETI Technische Instrumente GmbH. The camera has an integrated filter wheel, which is equipped with $v(\lambda)$, $x_{\text{short}}(\lambda)$, $x_{\text{long}}(\lambda)$ and $\bar{z}(\lambda)$ matching functions as well as an infrared and a glass filter, which has no significant spectrally selective absorption.

CHAPTER 3

SPECTRAL RAY FILES

In this chapter, the concept of the 6-dimensional data set "spectral ray file" is briefly described and motivated. Subsequently, ideal properties of measurement techniques to obtain spectral ray files are deduced and used to assess the briefly and solely qualitatively described state of the art.

3.1 SPECTRAL RAY FILES AND THE PLENOPTIC FUNCTION

The 7-dimensional plenoptic function $P(X, Y, Z, \phi, \vartheta, \lambda, t)$ is the most general light source description in the context of the ray optic assumption if polarization is neglected [60]. Each ray is described by its 3-dimensional starting point (X, Y, Z), the angular ray direction (ϕ, ϑ) and the wavelength λ. In the lighting industry, temporal variations t are usually neglected as the lighting distribution is assumed to be static or at least quasi-static, which reduces the plenoptic function to the 6-dimensional non-temporal plenoptic function $R(X, Y, Z, \phi, \vartheta, \lambda)$. If there is no spectral information available, the plenoptic function is reduced to the 5-dimensional usually photometric or pseudo-radiometric non-spectral-non-temporal plenoptic function $M(X, Y, Z, \phi, \vartheta)$.

A ray file is a discrete representation of the plenoptic function. Each directly measured ray file represents the 5-dimensional non-spectral-non-temporal plenoptic function because the spectral information is lost due to the optical integration on the ILMD (Imaging Luminance Measurement Device) as described in section 2.2 and 2.4.

In contrast, a spectral ray file represents one additional dimension, which is the wavelength as shown in Fig. 3.1 so that each ray has a specific starting point, angular direction and wavelength. Before the measurement properties and state-of-the-art techniques to create spectral ray files are discussed in section 3.3 and 3.4, section 3.2 explains the advantages of spectral ray files over a conventional ray file in commercial ray tracing software.

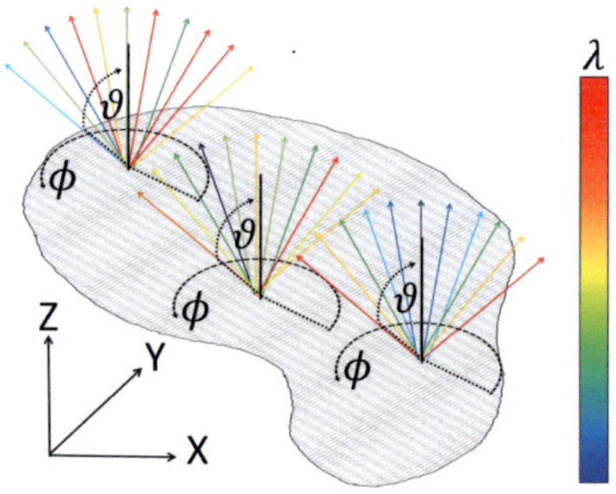

Figure 3.1: Schematic visualization of the discrete representation of the 6-dimensional plenoptic function.

3.2 APPLICATION OF SPECTRAL RAY FILES

According to the assumption of geometrical optics, the wavelength is unimportant during ray tracing and only the geometric relations of rays and surfaces as well as material parameters are relevant. The wavelength is only required to calculate photometric quantities, which depend on actinic action spectra. Examples are light and object colors, color rendering, blue light hazard or circadian effects. Thus, the wavelength does not seem to be necessary for the tracing process itself. However, this assumption is wrong because the wavelength does have an indirect influence on the ray tracing process.

If a ray hits a material interface, the ray tracer needs to assign the physical parameters of the materials such as absorption coefficient and the index of refraction. These parameters depend on the wavelength. For instance, in the visible part of the spectrum the index of refraction of classical optical materials is usually higher for smaller wavelengths. According to Snell's Law, this leads to different angles of refraction for different wavelengths, which is referred to as chromatic aberration. The image of the prism shown in Fig. 1.1 (b) in chapter 1 is an extreme example. This effect also occurs in the case of lenses and can lead to color shadows. As shown in Fig. 3.2 the cut-off line of automotive head lights may also show color shadows caused by chromatic aberration. The same can happen if different spectral absorptions occur in the system, which is for instance typical in rear lights.

This wavelength dependency is also the reason why it is not feasible to obtain any photometric quantity "prior" to the ray tracing. The effect of the wavelength dependent material parameters changes the results, as a function of angular or spatial dimension on the sensors of the ray tracer as it has in Fig. 3.2. Therefore, the design process of a complex optical system requires the inclusion of the wavelength dependent material parameters and the spectral information of the source.

In the case of conventional light sources such as tungsten halogen lamps, a global spectrum can be added to the ray file because the spectrum does not change as a function of spatial or angular dimension. The spectrum serves as a global probability density function which statistically assigns a wavelength to each ray. The optical design process of complex systems thus only requires modeling the wavelength dependence of the material parameters.

Figure 3.2: Color shadow caused by chromatic aberration at the cut-off line of an automotive head light.

However, in the case of LEDs there might be spectral variations as a function of angular or spatial dimension which can be caused by the spatial separation or by different intensity distributions of the individual LED chips and conversion phosphors. Then, attaching a global spectrum leads to errors [61]. To minimize these errors and achieve a realistic light source model, the conventional 5-dimensional ray file has to be replaced by a 6-dimensional spectral ray file. To sum up, to achieve precise spectral and colorimetric simulation results in commercial ray tracing software, realistic light source models incorporating the complete spectral information of the source are required.

3.3 PROPERTIES OF SPECTRAL RAY FILE MEASUREMENTS

The aim of computer-aided optical design is to accelerate the development process of optical components and optical systems. The spectral ray files should allow colorimetric and spectral simulations with realistic predictions within established non-sequential simulation software. Further, the measurement effort has to be reasonable with respect to the cost and time saving aspect of the computer aided lighting process. On the one hand it makes no sense to create a "perfect ray file" because there are other model assumptions in the ray tracer, such as manufacturing tolerances of the CAD models, which create a bias as well and because the required storage and measurement effort would increase significantly. On the other hand a fast and cheap measurement technique which does not provide useful results will not be used in any application. Finding the right trade-off between these aspects is crucial for each measurement task.

Spectral deviation The spectral deviation between two spectra may be described as residual sum of squares (RSS) or as chromaticity distance $\Delta u'v'$. While the RSS is not prone to metamerism and suited for comparative analyses, its non-interpretable values are not suited for assessing absolute deviations. In these cases, the interpretable chromaticity distance $\Delta u'v'$ is used. Typically, both figures of merit are highly correlated. Naturally, the tolerable spectral deviation of a spectral ray file always depends on the precise task. A typical one would be ensuring the color uniformity of an illumination. This is challenging because the human eye is a very sensitive detector when it comes to color discrimination. Nevertheless, the spectral deviation of a spectral ray file has to ensure at least the usability in such applications. The exact color discrimination threshold in $u'v'$ depends on the light color and the viewing conditions. Thus it varies from individual to individual and therefore also from study to study. In the case of white illumination, which has tough uniformity requirements, it typically lies in the range around $4 \times 10^{-3} < \Delta u'v' < 1.6 \times 10^{-2}$ in practical applications [62].

Memory requirements and interface The generated data format has to be capable of being integrated in existing non-sequential ray tracing software. Even if a measurement technique creates adequate results in a short amount of time, the results have to retain their properties in the established software without a complex adaption of the simulation software. Otherwise the usability decreases. The memory requirements should also remain reasonable because ray files are often provided on the home page of the LED manufacturers for download. This may become critical in the case of a 6-dimensional quantity. In the best case a standardized format as TM-25 [17], [55] can be used directly.

Measurement system As mentioned above the measurement effort has to be reasonable. Measuring the 5-dimensional quantity ray file with a moving scan typically requires between several minutes and a few hours. The measurement time depends on the measurement resolution, scanning range and the integration times of the sensors. As these measurement times are currently tolerated, the measurement time of a spectral ray file should be in the same order of magnitude. In particular, a start-stop-motion of the goniometer should be avoided because it heavily increases the measurement time. Another aspect that should be considered is the potential need for additional non-conventional measurement equipment such as a hyperspectral camera. It is a structural and financial advantage, if a method only requires established measurement equipment.

3.4 STATE-OF-THE-ART METHODS TO CREATE SPECTRAL RAY FILES

The most obvious and simple method to create spectral ray files would be an individual ILMD-based measurement of each LED source in the system followed by their combination. However, this approach has several problems [61].

First, the individual sources influence each other. One example is the thermal induced temperature drift, which is caused by unavoidable thermal losses of neighboring LEDs. The thermal drift affects the spectral shapes and intensities and therefore light color, overall flux and flux relations of the different LEDs. Second, not all LED systems allow the different chips or chip colors to be powered individually. Finally, in the case of phosphors an individual measurement is not physically possible because the phosphor is powered by LED radiation [61].

Consequently, over the last years several different measurement methods to create spectral ray files have been proposed. Most of them rely on conventional measurement techniques. They combine ILMD-based measurements with spectral measurements. The spatially resolved ILMD-based measurements are required to be goniometric in order to include angularly resolved information.

The spectral information of an ILMD-based measurement is lost due to the optical integration of the spectrum on the ILMD sensor. In combination with a goniometer it provides the 5-dimensional $M(X, Y, Z, \phi, \vartheta)$. However, it is possible to weight the spectrum by using different optical transmission filters prior to the integration on the chip. Each filter optically weights the spectrum differently and therefore generates different spectrally integrated pseudo-radiometric information. The number of required ILMD-based measurements with different optical filters varies in the state-of-art methods. On the other hand, the spectral measurements do not generate spatially resolved information because the spectroradiometer images one "pixel", but spectrally resolved information. In combination with the goniometric system the spectral information may be obtained as a function of the angle. Nevertheless, only the combination of both the ILMD and the spectroradiometer provides enough information in terms of spatial, angular and spectral resolution to create a spectral ray file, which represents

$R(X, Y, Z, \phi, \vartheta, \lambda)$, if no additional measurement equipment, such as hyperspectral cameras, are used.

Rykowksy summed up the four different possibilities for combining the spectral measurements with the ILMD-based measurements with respect to the required measurement effort [63]. Table 3.1 shows proposed state-of-the-art methods and their assignment to these four options. The columns distinguish the measurement methods between the need of angularly resolved spectral measurements and one spectral measurement. Note that only one state-of-the-art method avoids angularly resolved spectral measurements. The rows distinguish the measurement methods between the required number of ILMD-based measurements with different optical transmission filters, which is either equal to one or larger.

Table 3.1: Concepts to create spectral ray files $R(X, Y, Z, \phi, \vartheta, \lambda)$ with conventional measurement equipment and allocated state-of-the-art methods

		Spectral measurement	
		Global	**Angularly resolved**
ILMD-based measurements	One	*Spectral variations not accountable*	PCA method (3.4.3)
	Several	Blue/Yellow approach (3.4.4)	Spectral adjusting (3.4.1) Polychromatic ray data (3.4.2)

The option which combines only one ILMD-based measurement with one global spectral measurement does not contain any spectral information as a function of angular or spatial dimension. It can therefore not be used to create spectral ray files but only a ray file with a fixed spectrum. The three remaining options have all been used to create spectral ray files. They are evaluated in the subsections 3.4.1-3.4.4. The methods discussed in subsections 3.4.5 and 3.4.6 cannot be assigned to the table directly as they utilize different measurement approaches.

3.4.1 SPECTRAL ADJUSTING

One of the first methods to create spectral ray files was published by Radiant Vision in 2011 [63]. This method generally requires angularly resolved spectral measurements and ILMD-based measurements with at least three optical filters to obtain spatially resolved colorimetric information at the same angular positions. For each selected starting location (ILMD pixel) the associated far field spectrum is adjusted to match the measured colorimetric information. The adjustments are based on an iterative algorithm such that "as much features of the original spectral power distribution as possible remain" [63]. A more precise description of the algorithm is not provided as it is the intellectual property of Radiant Vision. The algorithm proceeds until the spectrum matches the measured color within a user defined tolerance. The authors suggest a colorimetric tolerance of $\triangle u'v' < 0.003$. The resulting data format provides an adjusted spectrum at each location and angle of radiation. This format cannot be used in a ray tracer directly because one ray requires one wavelength and not a complete spectrum. Therefore, a specific wavelength is assigned to each ray based on the provided spectrum of the starting point and the ray direction to allow the ray tracing. The memory requirements of the 6-dimensional raw data prior to the wavelength assignment are rather high.

The spectral adjustment can be prone to metamerism as the spectral adjustment is not unique since different spectra can produce the same color stimulus. This problem intensifies if the number of spectral sources increases and the term "spectral feature" becomes harder to define. The largest disadvantage, however, is the measurement time because angularly resolved spectral measurements are recommended to obtain good results. Nevertheless, this method can produce angularly and spectrally resolved spectral ray files by using only conventional measurement techniques.

3.4.2 Polychromatic ray data

The Opsira GmbH published another method to create spectral ray files [64]. It combines between seven and ten ILMD-based measurements with optical bandpass filters, which are evenly spaced across the visible spectrum, to create the same amount of different ray files. This ray file set serves as the basis for the ray tracing simulation [61], [65]. An angularly resolved spectral measurement allows the absolute radiometric weighting of each ray file by integrating the measured spectra in the defined spectral intervals and over the angular dimension. The authors note that strong spectral variations along the angular measurements may require an individual weighting of each ILMD picture prior to the ray generation in order to improve the absolute weighting [61]. This requires the angular positions of the ILMD-based measurements to match those of the spectral measurements.

The spectral deviations as well as the memory requirements and the measurement time depend on the number of filters used to create the ray file set. However, the number of ILMD-based measurements is always comparably high and the required angularly resolved spectral measurements result in a start-stop-motion of the goniometer. Both facts lead to a rather high measurement time. Nevertheless, the obtained set of ray files can be used directly in ray tracing software and produces angularly and spatially resolved spectral information. Although the whole concept relies only on conventional measurement equipment, it is the most general concept so far as it works for all kinds of stable light sources and not just all kinds of LEDs.

3.4.3 Principal component analysis

Jacobs et. al. proposed a method to create spectral ray files which contain brightness variations as a function of angular and spatial dimension and spectral variations as a function of angular dimension [66]. This approach is the only method requiring just one ILMD-based measurement but still requiring angularly resolved spectral measurements.

The angularly resolved spectral measurements are used to deduce a set of principal components whose weighted sum is capable of reconstructing the measured spectra. Afterwards the weighting factors are obtained at each angular spectral measurement position and used to interpolate the weighting factors at all required angular positions. This information is then combined with the angularly and spatially resolved information of the ILMD-based measurement [66]. The number of principal components used to reconstruct the spectra influences both spectral deviations and memory requirements. However, since the principal components are based on a PCA, the relative importance of additional basis functions decreases strongly and their number can remain quite small.

Advantages include the data compression as well as the requirement of solely conventional measurement equipment. An even larger advantage is the measurement time compared to the approaches discussed in subsection 3.4.1 and 3.4.2. Although the approach still requires angularly resolved spectral measurements, the measurement time is comparably low because the angularly resolved spectral measurements do not have to occur necessarily at the same positions as the ILMD-based measurements due to the possible interpolation. This is especially true in the case of rotationally symmetric spectral variations because only the polar angle requires an angular resolution. In the non-rationally symmetric case, the angularly resolved measurements still lead to high measurement times. The precision of the PCA and the interpolation increases in the case of additional spectral measurements. A large disadvantage is that the spectral variation is only described as function of the angular dimension. Therefore, this approach is more suited to describe spatially uniform sources such as OLEDs (organic light emitting diodes) but lacks precision in the case of spatially varying spectral information.

3.4.4 BLUE/YELLOW APPROACH

The LED manufacturer Osram has proposed the only measurement technique which requires a single spectral measurement in addition to two ILMD-based measurements. This approach is mainly used to describe phosphor converted white LEDs and is typically referred to as the "Blue/Yellow" approach or the "Blue/Yellow" ray files [16], [67]. The main idea of this method is to separate a global measured spectrum into two parts, which represent the physical sources blue LED and yellow phosphor. ILMD-based measurements with a blue optical filter and a yellow optical filter are performed to create two ray files. The spectrum is separated at a specific wavelength into a blue and yellow part as well. Each ray file is assigned to its sharp spectral part describing its relative change as a function of angular and spatial dimension. The relative weighting between the ray files as well as the absolute weighting with respect to the radiometric flux is considered by a global weighting factor for each ray file. Although the procedure to obtain these two factors is not described in [16], [67], it is fair to assume that these weightings are based on the global measured spectrum and the overall flux of the LED. A similar method is described in subsection 4.2.2.

The main advantage of this approach is the short measurement time because there is only a need for two ILMD-based measurements and no angularly resolved spectral measurement. As the angularly and spatially resolved spectral information is provided by two ray files, it is directly available in the ray tracer and has only small memory requirements. In addition it is compatible to the standardized ray file format IES TM-25.

However, one has to accept the introduction of a systematic error if this method is used. Although the spectrum consists of two spectral sources, which are the physical LED and phosphor spectrum, there is a difference between the physical spectra and the assigned spectral parts. The size of the error depends on the chosen separation wavelength, the characteristics of the source, the global spectrum and also on the actual filters used to create the ray files. If the filters transmit only a spectral range, which represents one spectral source directly without overlap to another source, the resulting ray files have the correct relative distribution.

3.4.5 HYPERSPECTRAL MEASUREMENTS

Lee et. al proposed a measurement technique which combines angularly resolved spectral measurements with one hyperspectral image [68]. First, all measurements are used to deduce sharply separated spectral parts. This is done based on a correlation analysis of the spectral change in the near- and far-field. Each spectral band is treated as an individual source.

The hyperspectral image provides the information regarding the 2-dimensional starting points. The angularly resolved spectral measurements serve as a data base of the ray directions for each spectral band. The 2-dimensional starting points on an emitting plane as well as the directions are determined by a Monte Carlo approach for each source. A ray tracer is used to compare simulation results of the model which consists of the sum of all sources with additional photometric and colorimetric mid-field measurement data. The mid-field is defined according to [69]. It is the region between the near field (vector region) and the far field (Fraunhaufer region) in which the light pattern varies from a distance to another.

In general, the third dimension of the starting point is neither the same for different spectral bands nor provided by the geometrical 2-dimensional hyperspectral image. Therefore, the location of the emitting plane of each spectral band is also an optimization value. This is for instance necessary in the case of phosphor converted white LEDs because the phosphor usually 3-dimensionally encapsulates the blue LED. The starting points, ray directions and emitting plane locations are altered until the normalized cross-correlation (NCC) of the simulation result and the mid-field measurement data is above a user defined threshold. The authors recommend a NCC above 0.99.

This method does not require a conventional ILMD-based goniophotometric measurement. However, although it uses hyperspectral imaging, angularly resolved spectral measurements are still required and thus lead to a high measurement time. In addition, the mid-field region has to be measured and the separation into sharp spectral bands differs from the true physical basis spectra as described in subsection 3.4.4.

3.4.6 Physical LED models

The approach discussed in this subsection will only be mentioned peripherally as its main purpose is not the generation of spectral ray files but the analysis and improvement of LED sources and phosphor parameters [70]–[72]. The idea is to create a complete geometrical and optical physical model of the LED and, if present, the phosphor within the ray tracer.

The modeling starts with the creation of a precise LED-CAD model, which includes for instance the electrode structures and encapsulation geometries. The modeling also requires the inclusion of additional measurement data such as for instance the blue LED emission spectrum and the phosphor absorption and re-emission spectra in order to correctly predict the wavelength conversion of the blue light in the phosphor matrix. While a semiconductor is optically described as Lambertian surface radiator, the scattering in the phosphor is modeled using Mie theory and requires additional parameters that have to be measured [73] or modeled [71]. In the case of [71], unknown intrinsic phosphor parameters within the phosphor matrix are altered until simulation results in ray tracing software are in agreement with additional photometric, colorimetric and/or spectral measurement data.

Compared to the spectral ray files described from subsection 3.4.1 to 3.4.5, the resulting models are not just descriptive with respect to the angularly and spatially resolved spectral information, but offer a deeper understanding of the LED source itself. However, the individual modeling effort and required data amount and therefore the overall creation time is larger than in the other methods. Nevertheless, if the complex models exist they can be used in a similar way to spectral ray files [74].

3.4.7 SUMMARY

Several state-of-the-art methods exist for the creation of spectral ray files. Most of them focus on the angular color uniformity of phosphor converted white LED. The fastest amongst them is the Blue/Yellow approach because it requires no angularly resolved spectral measurements. The two ILMD-based measurements provide a high angular and spatial resolution. The sharp spectral separation is its main limitation, which is also true in the case of the hyperspectral measurements. The PCA approach can also be used to obtain a spectral ray file of a solely angularly varying source. In contrast to the Blue/Yellow approach, it is not limited to LEDs and was successfully applied on a white OLED. It requires only one ILMD-based measurement, angularly resolved spectral measurements at a comparable low resolution and the assumption that the source has no spatially spectral variations.

To obtain a spectral ray file of a spatially and angularly varying source either the spectral adjusting or the polychromatic ray data can be used. However, both methods require time-consuming angularly resolved spectral measurements at the positions of the ILMD-based measurements. Although the spectral adjusting is usually faster than the polychromatic ray data because it utilizes only three ILMD-based measurements, it is also prone to metamerism and cannot be used for complex LED systems such as an RGBW LED. The only remaining measurement method besides setting up a physical model, is thus the polychromatic ray data. However, a spectral ray file of a Red/White LED system required eight ILMD-based measurements in addition to the angularly resolved spectral measurements [61]. As with the spectral adjusting, the measurement time can be reduced by decreasing the angular resolution. A technique which is both general and comparably fast does not exist until today.

CHAPTER 4

SPECTRAL RAY FILES BASED ON PHYSICALLY MOTIVATED BASIS SPECTRA (PMBS)

The aim of this chapter is to introduce the general concept of the measurement method. Starting from the direct generalization of the Blue/Yellow approach introduced in subsection 3.4.4, the physically motivated generalization is deduced. According to the main assumption that all spectral variations of the plenoptic function can be described as the weighted sum of physical basis spectra, it is named "physically motivated basis spectra" or PMBS. This chapter includes all fundamental assumptions, mathematical descriptions, and the principal limitations as well as a first outline of the complete workflow, which serves as starting points for chapters 5-7, which describe the practical application of the main steps of PMBS.

4.1 DIRECT GENERALIZATION OF THE BLUE/YELLOW APPROACH

The plenoptic function $R(X, Y, Z, \phi, \vartheta, \lambda)$ describes the radiation of a source in terms of starting point (X, Y, Z), direction (ϕ, ϑ) and wavelength λ if temporal variations and polarization can be neglected. It is therefore the most complete description of a radiation source if the ray optic assumption is used. Although the state-of-the-art methods described in section 3.4 are capable of providing this information in form of spectral ray files, they either require time-consuming angularly resolved spectral measurements or work only for phosphor converted white LEDs. Nevertheless, the state-of-the-art-methods provide very useful ideas and concepts, which are:

1. Spectral compression by the deduction of global basis spectra in order to reduce the storage amount as in the PCA approach (3.4.3), the Blue/Yellow approach (3.4.4) and the hyperspectral measurements (3.4.5)
2. Associating one ray file to each basis spectrum in order to use the spectral ray files directly in a state-of-the-art ray tracer as in the polychromatic ray data (3.4.2), the Blue/Yellow approach (3.4.4) and the hyperspectral measurements (3.4.5)
3. Incorporation of physical/prior knowledge of the device-under-test (DUT) as in the Blue/Yellow approach (3.4.4) and the physical LED modeling (3.4.6) instead of many (additional) redundant measurements
4. Checking the model based on the conducted and (or) additional measurements as in the hyperspectral measurements (3.4.5)
5. Avoidance of angularly resolved spectral measurements as in the Blue/Yellow approach (3.4.4) or at least a significant reduction as in the PCA approach to reduce the measurement time (3.4.3)

Taking a closer look, the Blue/Yellow approach from subsection 3.4.4 combines most principles. Its main drawback is that it is only described for phosphor converted white LEDs. Therefore, it has to be generalized.

The idea of the Blue/Yellow approach is to model each physical spectral source individually. The overall plenoptic function $R(X, Y, Z, \phi, \vartheta, \lambda)$ is composed of the sum of two individual spectrally separated plenoptic functions such that

$$R(X,Y,Z,\phi,\vartheta,\lambda) = R_{\mathrm{SB}}(X,Y,Z,\phi,\vartheta,\lambda) + R_{\mathrm{SY}}(X,Y,Z,\phi,\vartheta,\lambda) \quad (4.1)$$

in which $R_{\mathrm{SB}}(X,Y,Z,\phi,\vartheta,\lambda)$ is the plenoptic function of the blue part, which models the LED, and $R_{\mathrm{SY}}(X,Y,Z,\phi,\vartheta,\lambda)$ the plenoptic function of the yellow part, which models the phosphor. It is assumed that each individual plenoptic function can be divided into a non-spectral blue part $M_{\mathrm{A,SB}}(X,Y,Z,\phi,\vartheta)$ and a non-spectral yellow part $M_{\mathrm{A,SY}}(X,Y,Z,\phi,\vartheta)$ as well as an individual globally separated spectrum of the blue part $S_{\mathrm{SB}}(\lambda)$ and the yellow part $S_{\mathrm{SY}}(\lambda)$:

$$\begin{aligned} R_{\mathrm{SB}}(X,Y,Z,\phi,\vartheta,\lambda) &= M_{\mathrm{A,SB}}(X,Y,Z,\phi,\vartheta) \times S_{\mathrm{SB}}(\lambda) \\ R_{\mathrm{SY}}(X,Y,Z,\phi,\vartheta,\lambda) &= M_{\mathrm{A,SY}}(X,Y,Z,\phi,\vartheta) \times S_{\mathrm{SY}}(\lambda) \end{aligned} \quad (4.2)$$

The overall spectrum $S(\lambda)$ can be written as

$$S(\lambda) = S_{\mathrm{SB}}(\lambda) + S_{\mathrm{SY}}(\lambda) \quad (4.3)$$

with

$$\begin{aligned} S_{\mathrm{SB}}(\lambda) &= \begin{cases} S(\lambda) & 0 \leq \lambda < \lambda_{\mathrm{Cut}} \\ 0 & \lambda_{\mathrm{Cut}} \leq \lambda < \infty \end{cases}, \\ S_{\mathrm{YB}}(\lambda) &= \begin{cases} 0 & 0 \leq \lambda < \lambda_{\mathrm{Cut}} \\ S(\lambda) & \lambda_{\mathrm{Cut}} \leq \lambda < \infty \end{cases} \end{aligned} \quad (4.4)$$

where λ_{Cut} is the separation wavelength. The spectral separation is therefore a sharp spectral separation. The principle is visualized in Fig. 4.1. It shows a relative spectral distribution of a phosphor converted white LED, which is separated at a separation wavelength λ_{Cut} according to Eq. 4.4.

Individual global weightings for the plenoptic functions of the blue LED W_{SB} and the yellow phosphor W_{SY} are necessary, because the plenoptic function $M_n(X,Y,Z,\phi,\vartheta)$ is represented by a measured ray file, which is a relative distribution. The weightings ensure the correct relative relation between the individual plenoptic functions.

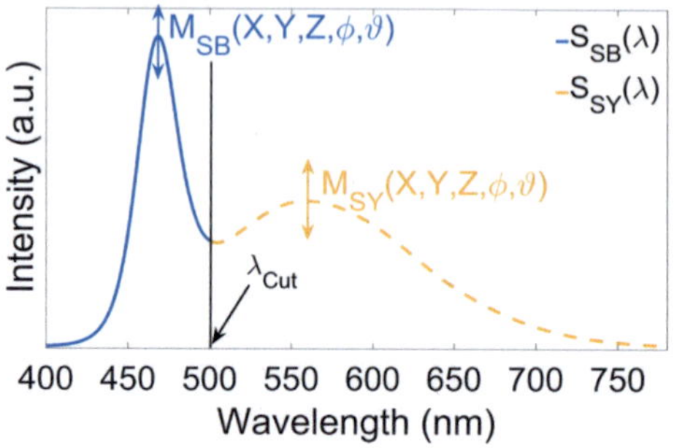

Figure 4.1: Principle of the Blue/Yellow approach.

If the weightings are absolute weightings, the combination of the plenoptic functions and spectra provides the correct radiometric flux. Then the absolute plenoptic functions $M_{\mathrm{A,SB}}(X, Y, Z, \phi, \vartheta)$ and $M_{\mathrm{A,SY}}(X, Y, Z, \phi, \vartheta)$ are defined as

$$\begin{aligned} M_{\mathrm{A,SB}}(X, Y, Z, \phi, \vartheta) &= W_{\mathrm{SB}} \times M_{\mathrm{SB}}(X, Y, Z, \phi, \vartheta) \\ M_{\mathrm{A,SY}}(X, Y, Z, \phi, \vartheta) &= W_{\mathrm{SY}} \times M_{\mathrm{SY}}(X, Y, Z, \phi, \vartheta) \end{aligned} \tag{4.5}$$

The absolute $M_{\mathrm{A,SB}}(X, Y, Z, \phi, \vartheta)$ and $M_{\mathrm{A,SY}}(X, Y, Z, \phi, \vartheta)$ provide the weighting as a function of spatial and angular dimension.

The direct generalization for n_{max} different sharply separated spectral sources $S_{S,n}(\lambda)$ would be

$$S(\lambda) = \sum_{n=1}^{n_{\mathrm{max}}} S_{\mathrm{S},n}(\lambda) \tag{4.6}$$

with

$$S_{\mathrm{S},n}(\lambda) = \begin{cases} 0 & \lambda < \lambda_{\mathrm{Cut},n-1} \\ S(\lambda) & \lambda_{\mathrm{Cut},n-1} \leq \lambda < \lambda_{\mathrm{Cut},n} \\ 0 & \lambda_{\mathrm{Cut},n} \leq \lambda < \infty \end{cases} \tag{4.7}$$

and $\lambda_{\mathrm{Cut},0} = 0$ nm. The variable $\lambda_{\mathrm{Cut},n}$ is the separation wavelength between the nth and nth+1 sharp spectral band. Each nth

spectral band has to be measured individually with a suitable filter to obtain $M_{S,n}(X,Y,Z,\phi,\vartheta,\lambda)$. Afterwards, the plenoptic function $R(X,Y,Z,\phi,\vartheta,\lambda)$ is the weighted sum of these ILMD-based goniophotometric measurements and the sharply separated spectra as described in Eq. 4.8.

$$\begin{aligned} R(X,Y,Z,\phi,\vartheta,\lambda) &= \sum_{n=1}^{n_{\max}} R_{S,n}(X,Y,Z,\phi,\vartheta,\lambda) \\ &= \sum_{n=1}^{n_{\max}} W_{S,n} \times M_{S,n}(X,Y,Z,\phi,\vartheta) \times S_{S,n}(\lambda) \end{aligned} \tag{4.8}$$

The deduction of the weighting factors $W_{S,n}$ is similar to those described in subsection 4.2.2 and is therefore not described in detail at this point.

Although Eq. 4.6 to Eq. 4.8 are general, their combination is not physically valid for each LED spectrum and the associated error can become large for certain LED spectra. The problem occurs if a sharp spectral distribution $S_{S,n}(\lambda)$ changes as a function of angular and spatial dimension and therefore the association of the usually pseudo-radiometric plenoptic function $M_n(X,Y,Z,\phi,\vartheta)$ to a sharply separated spectrum $S_{S,n}(\lambda)$ becomes erroneous. Sometimes there is a full spectral overlap as in the case of the combination of phosphor spectra and red or green LED, which is typical to enhance the color rendering of a luminaire. Then the generalization provided by Eq. 4.6 to Eq. 4.8 is not valid due to the non valid assumption of constant sharp separated basis spectra.

The problem of spectral overlap is visualized in Fig. 4.2, which shows three measured spectra at different angular positions from an LED system, which consists of phosphor converted white LEDs and additional red LEDs. The sharp spectral separation at $\lambda_{\text{Cut},1}$ is reasonable because the relative spectra $S_{S,1}(\lambda)$ and the left hand side of the $S_{S,2}(\lambda)$ are similar. However, the sharp spectral separation into a red and a yellow part at $\lambda_{\text{Cut},2}$ results in a different relative spectral distribution

$S_{S,3}(\lambda)$ because no constant factor exists between the $S_{S,3}(\lambda)$ at the different positions. This violates the assumption in Eq. 4.8, which states that all $S_{S,n}(\lambda)$ do not depend on the angle of radiation. The change is caused by the large spectral overlap of the physical phosphor spectrum and the spectrum of the red LED. An experimental analysis of this problem in the case of a warm white phosphor converted LED is provided in section 8.1.

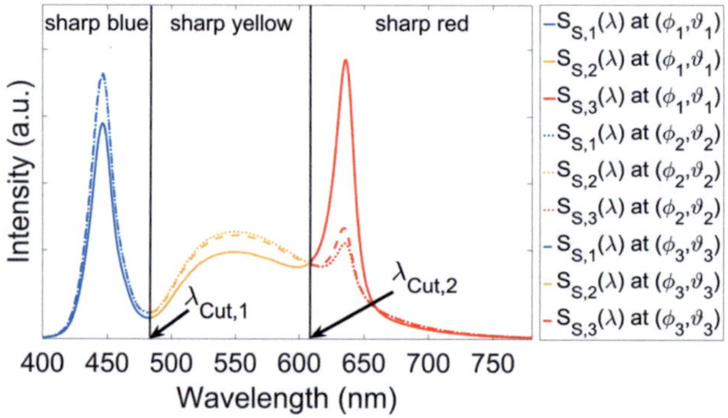

Figure 4.2: Problem of the direct generalization of the Blue/Yellow approach: The relative spectral distributions depend on the angular position.

To avoid this problem, a valid and more physical generalization of the Blue/Yellow approach is necessary. This generalization aims to avoid the less precise and arbitrary sharp spectral separation and leads to the physically motivated basis spectra approach.

4.2 PHYSICALLY MOTIVATED BASIS SPECTRA (PMBS)

To show the differences between the arbitrary and physical generalization of the Blue/Yellow approach, the physically relations of a single phosphor converted white LED will be formulated in equations similar to those in section 4.1. The physical realization of Eq. 4.3 uses the physical spectra of each source and describes the overall spectrum $S(\lambda)$ as the sum of the blue LED spectrum $S_{\text{Blue}}(\lambda)$ and the yellow phosphor spectrum $S_{\text{Phosphor}}(\lambda)$.

$$S(\lambda) = S_{\text{Blue}}(\lambda) + S_{\text{Phosphor}}(\lambda) \tag{4.9}$$

The main difference between Eq. 4.3 and Eq. 4.9 is that the spectra $S_{\text{Blue}}(\lambda)$ and $S_{\text{Phosphor}}(\lambda)$ in Eq. 4.9 are the physical spectra of the individual sources. Then the assumption that the relative spectrum of each individual physical radiation source is constant over the spatial and angular dimension becomes valid. This means it is not erroneous to assign a plenoptic function $A_n(X, Y, Z, \phi, \vartheta)$ to the physical basis spectrum $S_n(\lambda)$ for the blue LED and the phosphor such that

$$\begin{aligned} R(X, Y, Z, \phi, \vartheta, \lambda) = {} & A_{\text{Blue}}(X, Y, Z, \phi, \vartheta) \times S_{\text{Blue}}(\lambda) \\ & + A_{\text{Phosphor}}(X, Y, Z, \phi, \vartheta) \times S_{\text{Phosphor}}(\lambda) \end{aligned} \tag{4.10}$$

This is visualized in Fig. 4.3. At first glance, Eq. 4.1, 4.2 and Fig. 4.3 seem to be identical to Eq. 4.8 and Fig. 4.1. However, it is important to note that $A_n(X, Y, Z, \phi, \vartheta) \not\propto M_n(X, Y, Z, \phi, \vartheta)$. $M_n(X, Y, Z, \phi, \vartheta)$ is the direct result of an ILMD-based measurement and thus a usually pseudo-radiometric quantity of all sources. The radiometric distribution $A_n(X, Y, Z, \phi, \vartheta)$ in contrast is always assigned to precisely one individual source and not necessarily proportional to any direct measurement result.

The generalization for LED systems with n_{max} different spectral sources $S_n(\lambda)$ would be similar to Eq. 4.6 and Eq. 4.8:

$$S(\lambda) = \sum_{n=1}^{n_{\text{max}}} S_n(\lambda) \tag{4.11}$$

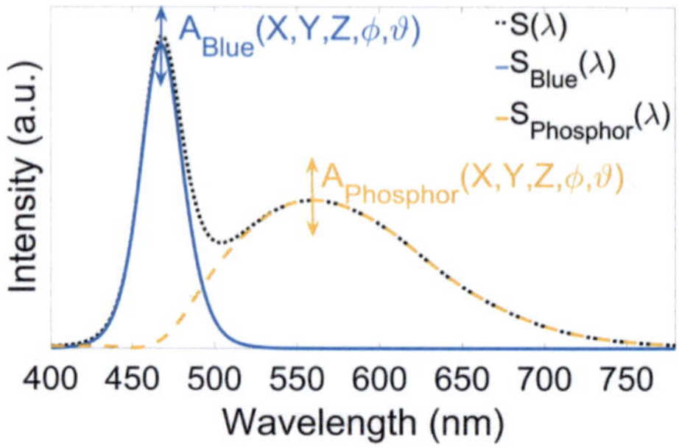

Figure 4.3: Principle of PMBS approach in the case of a phosphor converted white LED.

The plenoptic function $R(X, Y, Z, \phi, \vartheta, \lambda)$ is described as the weighted sum of the $n_{\max}$ physical basis spectra $S_n(\lambda)$, which exists in the DUT:

$$R(X, Y, Z, \phi, \vartheta, \lambda) = \sum_{n=1}^{n_{\max}} A_n(X, Y, Z, \phi, \vartheta) \times S_n(\lambda) \tag{4.12}$$

This is one of the most important equations in this chapter. It means that although the plenoptic function $R(X, Y, Z, \phi, \vartheta, \lambda)$ varies spectrally as a function of the angular and spatial dimension, its underlying relative spectral distributions are constant. $R(X, Y, Z, \phi, \vartheta, \lambda)$ can therefore be described by the weighted sum of physical basis spectra, which occur in the system. All spatial and angular variations are described by the relative radiometric $A_n(X, Y, Z, \phi, \vartheta)$, which change as a function of angular and spatial dimension.

However, one problem remains. As previously mentioned, a direct ILMD-based measurement of the relative radiometric distributions $A_n(X, Y, Z, \phi, \vartheta)$ is neither assumed nor generally possible. To be more precise, there is no optical filter which can measure only the relative change of one basis spectrum, if that basis spectrum is fully overlapped by other basis spectra. The relative radiometric distributions $A_n(X, Y, Z, \phi, \vartheta)$ have to be obtained using the combination of different pseudo-radiometric ILMD-based measurements. The procedure is explained in the next subsection.

4.2.1 SPECTRAL RECONSTRUCTION MATRIX

The absolute $M_{A,n}(X,Y,Z,\phi,\vartheta)$ obtained with the ILMD-based measurements depends on the effective spectral sensitivity $s_{\tau,\mathrm{eff},n}(\lambda)$ of the ILMD-system. According to Eq. 4.13 $s_{\tau,\mathrm{eff},n}(\lambda)$ is the product of the transmission function of the actual optical filter $\tau_n(\lambda)$ used to perform the measurement, the optional transmission function of a neutral density filter $\tau_{\mathrm{ND},n}(\lambda)$ and the spectral system sensitivity function $s_{\mathrm{Sys}}(\lambda)$. The spectral system transmission function usually remains constant and includes the sensitivity of the optoelectronic sensor and the transmission of the lens optics

$$s_{\tau,\mathrm{eff},n}(\lambda) = \tau_n(\lambda) \times \tau_{\mathrm{ND},n}(\lambda) \times s_{\mathrm{Sys}}(\lambda). \tag{4.13}$$

The absolute measurement signal $M_A(X,Y,Z,\phi,\vartheta)$ of an ILMD-based measurement with the effective spectral sensitivity $s_{\tau,\mathrm{eff}}(\lambda)$ is given by

$$M_A(X,Y,Z,\phi,\vartheta) = \sum_{n=1}^{n_{\max}} A_n(X,Y,Z,\phi,\vartheta) \times \int S_n(\lambda) s_{\tau,\mathrm{eff}}(\lambda) d\lambda. \tag{4.14}$$

It is assumed that besides the measured signal $M_A(X,Y,Z,\phi,\vartheta)$ the relative basis spectra $S_n(\lambda)$ as well as all spectral transmissions and sensitivities and therefore $s_{\tau,\mathrm{eff}}(\lambda)$ are known. This means that only the $n_{\max}$ $A_n(X,Y,Z,\phi,\vartheta)$ are unknown variables in Eq. 4.14. If m different optical filters are used, they produce m different effective spectral sensitivities $s_{\tau,\mathrm{eff},m}(\lambda)$ and thus m spectrally weighted pseudo-radiometric measurement values $M_{A,m}(X,Y,Z,\phi,\vartheta)$, which are given by

$$\begin{aligned} M_{A,1}(X,Y,Z,\phi,\vartheta) &= \sum_{n=1}^{n_{\max}} A_n(X,Y,Z,\phi,\vartheta) \times \int S_n(\lambda) s_{\tau,\mathrm{eff},1}(\lambda) d\lambda \\ &\dots \\ M_{A,m}(X,Y,Z,\phi,\vartheta) &= \sum_{n=1}^{n_{\max}} A_n(X,Y,Z,\phi,\vartheta) \times \int S_n(\lambda) s_{\tau,\mathrm{eff},m}(\lambda) d\lambda. \end{aligned} \tag{4.15}$$

These equations are a system of linear equations, which has a unique solution if the number of different optical filter measurements equals the number of physical basis spectra $m = n_{\max}$. Equation 4.15 can be rewritten in its matrix form as

$$\begin{bmatrix} M_{\mathrm{A},1}(X,Y,Z,\phi,\vartheta) \\ \vdots \\ M_{\mathrm{A},n_{\max}}(X,Y,Z,\phi,\vartheta) \end{bmatrix} = M_{\mathrm{S}\mathcal{T}} \cdot \begin{bmatrix} A_1(X,Y,Z,\phi,\vartheta) \\ \vdots \\ A_{n_{\max}}(X,Y,Z,\phi,\vartheta) \end{bmatrix}, \tag{4.16}$$

with

$$M_{\mathrm{S}\mathcal{T}} = \begin{bmatrix} \int S_1(\lambda) s_{\mathcal{T},\mathrm{eff},1}(\lambda) d\lambda & \cdots & \int S_{n_{\max}}(\lambda) s_{\mathcal{T},\mathrm{eff},1}(\lambda) d\lambda \\ \vdots & \ddots & \vdots \\ \int S_1(\lambda) s_{\mathcal{T},\mathrm{eff},n_{\max}}(\lambda) d\lambda & \cdots & \int S_{n_{\max}}(\lambda) s_{\mathcal{T},\mathrm{eff},n_{\max}}(\lambda) d\lambda \end{bmatrix}$$

The matrix $M_{\mathrm{S}\mathcal{T}}$ summarizes all available information about the effective spectral sensitivity of the measurement system and the physical basis spectra of the DUT. Equation 4.16 therefore relates the absolute pseudo-radiometric measurement values $M_{\mathrm{A},n}(X,Y,Z,\phi,\vartheta)$ to the unknown radiometric amplitudes $A_n(X,Y,Z,\phi,\vartheta)$, which will be assigned to the $n_{\max}$ physical basis spectra. The radiometric distributions $A_n(X,Y,Z,\phi,\vartheta)$ can be reconstructed by inverting the matrix $M_{\mathrm{S}\mathcal{T}}$. The reconstruction equation is given as

$$\begin{bmatrix} A_1(X,Y,Z,\phi,\vartheta) \\ \vdots \\ A_{n_{\max}}(X,Y,Z,\phi,\vartheta) \end{bmatrix} = M_{\mathrm{S}\mathcal{T}}^{-1} \cdot \begin{bmatrix} M_{\mathrm{A},1}(X,Y,Z,\phi,\vartheta) \\ \vdots \\ M_{\mathrm{A},n_{\max}}(X,Y,Z,\phi,\vartheta) \end{bmatrix} \tag{4.17}$$

The matrix $M_{\mathrm{S}\mathcal{T}}^{-1}$ will be called the spectral reconstruction matrix. The reconstructed distributions can be used to calculate the plenoptic function and thus to create the spectral ray file. The general procedure is summed up in the following subsection.

4.2.2 RELATIVE AND ABSOLUTE WEIGHTINGS

It is important to note that Eq. 4.14 til 4.17 are only valid if absolute values are used. However, in this context the $M_n(X, Y, Z, \phi, \vartheta)$ are based on measured ray files. They are therefore provided as unweighted distributions. According to Eq. 4.5 the absolute distributions are given as

$$\begin{bmatrix} M_{A,1}(X,Y,Z,\phi,\vartheta) \\ \vdots \\ M_{An_{max}}(X,Y,Z,\phi,\vartheta) \end{bmatrix} = \begin{bmatrix} W_1 & \cdots & 0 \\ \vdots & \ddots & \vdots \\ 0 & \cdots & W_{n_{max}} \end{bmatrix} \cdot \begin{bmatrix} M_1(X,Y,Z,\phi,\vartheta) \\ \vdots \\ M_{n_{max}}(X,Y,Z,\phi,\vartheta) \end{bmatrix} \tag{4.18}$$

where the diagonal matrix of the weightings can be written as the product of the relative weighting matrix W_R and an absolute scaling factor W_A

$$\underbrace{\begin{bmatrix} W_1 & \cdots & 0 \\ \vdots & \ddots & \vdots \\ 0 & \cdots & W_{n_\mathrm{max}} \end{bmatrix}}_{\text{absolute weightings}} = W_\mathrm{A} \underbrace{\begin{bmatrix} W_{\mathrm{R},1} & \cdots & 0 \\ \vdots & \ddots & \vdots \\ 0 & \cdots & W_{\mathrm{R},n_\mathrm{max}} \end{bmatrix}}_{\text{relative weightings } W_\mathrm{R}} . \tag{4.19}$$

The separation of the absolute and relative weightings makes sense for two reasons. Firstly, the absolute scaling factor can be based on the photometer, which provides absolute luminous flux values with a smaller uncertainty than the ILMD. Secondly, the mathematical importance of both weightings differs. Only the relatives weightings are essential for PMBS. If Eq. 4.18 and 4.19 are inserted into Eq. 4.17 and the matrix multiplications are carried out the result is given by

$$
\begin{aligned}
A_1(X,Y,Z,\phi,\vartheta) &= W_{\mathrm{A}} \times \Big[M_1(X,Y,Z,\phi,\vartheta) \times M_{\mathrm{S}\tau}^{-1}(1,1) W_{\mathrm{R},1} \\
&\quad + \cdots \\
&\quad + M_{n_{\max}}(X,Y,Z,\phi,\vartheta) \times M_{\mathrm{S}\tau}^{-1}(1,n_{\max}) W_{\mathrm{R},n_{\max}} \Big] \\
&\ \ \vdots \\
A_{n_{\max}}(X,Y,Z,\phi,\vartheta) &= W_{\mathrm{A}} \times \Big[M_1(X,Y,Z,\phi,\vartheta) \times M_{\mathrm{S}\tau}^{-1}(n_{\max},1) W_{\mathrm{R},1} \\
&\quad + \cdots \\
&\quad + M_{n_{\max}}(X,Y,Z,\phi,\vartheta) \times M_{\mathrm{S}\tau}^{-1}(n_{\max},n_{\max}) W_{\mathrm{R}n_{\max}} \Big]
\end{aligned}
\tag{4.20}
$$

This shows that the relative weightings are necessary to perform the reconstruction because they effectively change the complete reconstruction matrix. If the relative weightings are ignored, the spectral reconstruction does not provide the correct radiometric distributions $A_n(X,Y,Z,\phi,\vartheta)$.

In contrast, the absolute scaling factor W_{A} can be factored out in Eq. 4.20 and therefore also remains as a global scaling factor, if the distributions $A_n(X,Y,Z,\phi,\vartheta)$ from Eq. 4.20 are inserted into Eq. 4.12. W_{A} is chosen so that it relates the integral of the plenoptic function to an absolute radiometric or photometric value.

The determination of the relative weighting factors $W_{\mathrm{R},n}$ is mandatory and described in the following. Two methods to obtain the correct relative weightings $W_{\mathrm{R},n}$ will be discussed. The first method uses measured values obtained directly from the goniophotometric ILMD-based measurements. The availability of these values depends on the measurements. The second method relates the relative distribution $M_n(X,Y,Z,\phi,\vartheta)$ to a spectral measurement in the known geometrical setup and deduces the weightings by using Eq. 4.16. Finally, the absolute weighting W_{A} can be applied to ensure the correct radiometric flux such that

$$\begin{bmatrix} M_{\mathrm{A},1}(X,Y,Z,\phi,\vartheta) \\ \vdots \\ M_{\mathrm{A},n_{\max}}(X,Y,Z,\phi,\vartheta) \end{bmatrix} = W_{\mathrm{A}} \times W_{\mathrm{R}} \cdot \begin{bmatrix} M_1(X,Y,Z,\phi,\vartheta) \\ \vdots \\ M_{n_{\max}}(X,Y,Z,\phi,\vartheta) \end{bmatrix}.$$

$$W_{\mathrm{R}} = \begin{bmatrix} W_{\mathrm{R},1} & \cdots & 0 \\ \vdots & \ddots & \vdots \\ 0 & \cdots & W_{\mathrm{R},n_{\max}} \end{bmatrix} \tag{4.21}$$

Furthermore, it should be mentioned that in contrast to the weightings of the measurement values, an arbitrary non-zero weighting of the basis spectra does not change the result in Eq. 4.12. The basis spectra weightings change the reconstructed $A_n(X,Y,Z,\phi,\vartheta)$ but also the initial weightings of $S_n(\lambda)$ and therefore not their product in the final Eq. 4.12. Mathematical details are provided in the Appendix A.1.

METHOD 1: CAMERA LUMINOUS FLUX

While the luminous flux of the ray file typically refers to the angular integrated photometer measurements, some near field goniophotometers sum up all preprocessed weighted pseudo-radiance images of the ILMD-based measurement. These values can be used directly as weightings for the corresponding ray file and therefore provide the weighting factor.[1] If these values are not available, the weighting can be estimated based on a spectral measurement using the second method.

---

[1] In the case of the RIGO801 used in this thesis, this value is provided as "camera luminous flux" in the measurement results. It is corrected by a calibration factor to generate a luminous flux under the assumption that the device under test has a spectral distribution according to standard illuminant A. Each filter has its own calibration factor. The weighting of the camera luminous flux has to be negated by dividing the camera luminous flux by the individual calibration factor of the filter.

METHOD 2: INVERSE RECONSTRUCTION

The second and more general option to obtain the weighting is to use the geometrical information of the spectral measurement and the measurement ray files used to obtain $M_n(X, Y, Z, \phi, \vartheta)$. Typically, the spectrum is measured with the goniometer in the same setup as all ILMD-based measurements at a known position (ϕ, ϑ, r). The variable r is the measurement distance. The measured spectrum $S_{\text{M}}(\lambda, \phi, \vartheta, r)$ at the known position is a linear combination of the basis spectra $S_n(\lambda)$ according to

$$S_{\text{M}}(\lambda, \phi, \vartheta, r) = \sum_{n=1}^{n_{\max}} A_n(\phi, \vartheta, r) \times S_n(\lambda). \tag{4.22}$$

The amplitudes $A_n(\phi, \vartheta, r) \in (0, \infty)$ for the known position can be estimated using an optimization algorithm, in which the amplitudes $A_n(\phi, \vartheta, r)$ are the optimization variables by minimizing the residual sum of squares $RSS\,(A_n(\phi, \vartheta, r))$ defined as

$$RSS\,(A_n(\phi, \vartheta, r)) = \sum_{\lambda} \left(S_{\text{M}}(\lambda, \phi, \vartheta, r) - \sum_{n=1}^{n_{\max}} A_n(\phi, \vartheta, r) \times S_n(\lambda) \right)^2. \tag{4.23}$$

The amplitudes $A_n(\phi, \vartheta, r)$ can be used to estimate the expected measurement values $M_{\text{A},n}(\phi, \vartheta, r)$ at the known position (ϕ, ϑ, r) using Eq. 4.16. The relative weighting of the measurement can then be obtained by ray tracing of the relative measurement ray files towards the known position (ϕ, ϑ, r).

Figure 4.4 visualizes the concept. Each ILMD-based measurement is represented as a different color in the figure. A sensor is placed at the known position (ϕ, ϑ, r) of the spectroradiometer. The measurement distance r and the diameter of the sensor are included in the ray tracing, which ensures that near field information is used. The diameter of the sensor element is equal to the diameter of the optical probe of the spectroradiometer. Each measurement ray file is traced individually to that sensor to obtain the ray number $M_n(\phi, \vartheta, r)$ at position (ϕ, ϑ, r). Finally, the weighting $W_{R,n}$ can be calculated by

$$W_{R,n} = \frac{M_{A,n}(\phi, \vartheta, r)}{M_n(\phi, \vartheta, r)} \times N_{Ray,n}, \tag{4.24}$$

in which $N_{Ray,n}$ is the number of rays in the nth measured ray file. Note that 4.24 requires that all rays in the ray file have a constant amplitude. To sum up, the relative weightings $W_{R,n}$ are determined such that a spectral reconstruction at a specific position (ϕ, ϑ, r) is capable of reconstructing the measured spectrum at the same position. It is based on the measured ray files and the spectral reconstruction matrix. Basically, the PMBS method is used inversely.

The spectral measurement should contain spectral data of all basis spectra if this method is used. Typically, the main radiance direction at $\phi = 0°$ and $\vartheta = 0°$ is a suitable position for the spectral measurement. If the spectral measurement is an absolute measurement, the relative weightings already include the correct absolute weighting and thus lead to $W_A \approx 1$. If the spectral measurement only provides a relative distribution $S_M(\lambda)$, an additional absolute weighting has to be calculated.

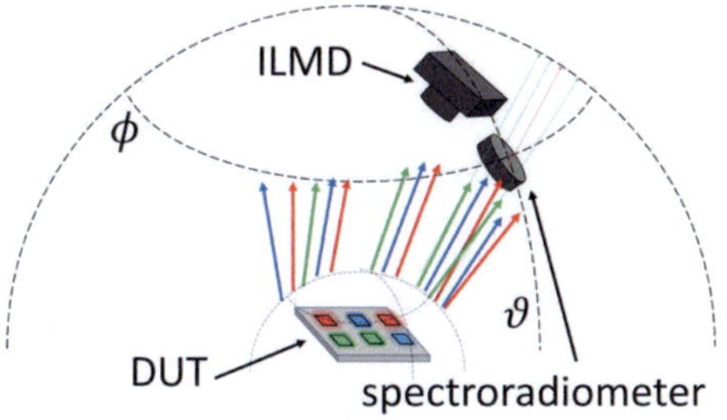

Figure 4.4: Deduction of relative weightings $W_{R,n}$: The rays of the different ILMD-based goniophotometric measurements (different ray colors) are traced towards the spectroradiometer and compared to a spectral measurement.

SPECIAL CASE: NORMALIZATION

The relative weightings can be replaced by a normalization of the spectral reconstruction matrix such that $M_{\mathrm{R},S\tau}$ is defined as,

$$\begin{bmatrix} 1 \\ \vdots \\ 1 \end{bmatrix} = M_{\mathrm{R},S\tau} \cdot \begin{bmatrix} 1 \\ \vdots \\ 1 \end{bmatrix} \tag{4.25}$$

and

$$M_{\mathrm{R},S\tau} = \begin{bmatrix} W_{\mathrm{norm},1} & \cdots & 0 \\ \vdots & \ddots & \vdots \\ 0 & \cdots & W_{\mathrm{norm},n_{\max}} \end{bmatrix} \cdot M_{S\tau} \tag{4.26}$$

resulting in

$$W_{\mathrm{norm},n} = \frac{1}{\sum_{m=1}^{n_{\max}} \int S_m(\lambda) s_{\tau,\mathrm{eff},n}(\lambda) d\lambda} \tag{4.27}$$

if two conditions are fulfilled:

1. The basis spectra deduction is based on a spectral measurement of the complete spectrum $S(\lambda)$.
2. All spectral sources follow the same relative distribution (for instance Lambertian) and have the same number of rays.

The first condition will be fulfilled in most cases because the spectral measurement is performed in the same goniometric setup as the ILMD-based measurements. The second one is the limiting condition but is for instance fulfilled in the case of RGB LEDs. Note that it is not fulfilled in the case of RGBW because the phosphor will have a different relative distribution.

It can be shown that this normalization equals the second method described above. The deduction of the amplitudes using the optimization in Eq. 4.23 results in $A_1 = 1 = \cdots = A_{n_{\max}}$ because of the first condition. Then Eq. 4.16 gives

$$M_{\mathrm{A},n} = \sum_{m=1}^{n_{\max}} \int S_m(\lambda) s_{\tau,\mathrm{eff},n}(\lambda) d\lambda \tag{4.28}$$

for all expected measurement values $M_{\mathrm{A},n}$ at the known position. The second condition provides that the ray tracing towards each position would result in the same $M_1 = \cdots = M_{n_{\max}}$. It is thus a constant for all measurement values and can be neglected such that the relative weightings of the measurements would be

$$W_{\mathrm{R},n} = \sum_{m=1}^{n_{\max}} \int S_m(\lambda) s_{\tau,\mathrm{eff},n}(\lambda) d\lambda \times N_{\mathrm{Ray},n} \tag{4.29}$$

for all measurements. If these weightings are inserted into Eq. 4.21 and then into Eq. 4.16, the result can be written as

$$W_A \times W_R \cdot \begin{bmatrix} M_1(X,Y,Z,\phi,\vartheta) \\ \vdots \\ M_{n_{max}}(X,Y,Z,\phi,\vartheta) \end{bmatrix} = M_{S\tau} \cdot \begin{bmatrix} A_1(X,Y,Z,\phi,\vartheta) \\ \vdots \\ A_{n_{max}}(X,Y,Z,\phi,\vartheta) \end{bmatrix}. \tag{4.30}$$

By multiplying the inverse of the weighting matrix from the left hand side the equation changes to (note that W_{A} is a scalar)

$$W_A \times \begin{bmatrix} M_1(X,Y,Z,\phi,\vartheta) \\ \vdots \\ M_{n_{max}}(X,Y,Z,\phi,\vartheta) \end{bmatrix} = W_R^{-1} \cdot M_{S\tau} \cdot \begin{bmatrix} A_1(X,Y,Z,\phi,\vartheta) \\ \vdots \\ A_{n_{max}}(X,Y,Z,\phi,\vartheta) \end{bmatrix}. \tag{4.31}$$

Since the weighting matrix is a diagonal matrix, the inverse is a diagonal matrix, in which each element is the inverted element of the original matrix. Thus it can be rewritten according to Eq. 4.26

$$W_A \times \begin{bmatrix} M_1(X,Y,Z,\phi,\vartheta) \\ \vdots \\ M_{n_{max}}(X,Y,Z,\phi,\vartheta) \end{bmatrix} = M_{R,S\tau} \cdot \begin{bmatrix} A_1(X,Y,Z,\phi,\vartheta) \\ \vdots \\ A_{n_{max}}(X,Y,Z,\phi,\vartheta) \end{bmatrix}. \tag{4.32}$$

This is a convenient result because it simplifies the procedure in some cases such as RGB LEDs.

ABSOLUTE WEIGHTING

The absolute weighting W_A can be based on the absolute photometric measurement, which is always performed during an ILMD-based goniophotometric measurement by using the photometer. The angularly integrated photometer values provide the luminous flux of the device under test. Another option is to use additional measurement equipment. Afterwards, the spectral ray files are weighted such that their luminous flux matches that of the absolute photometric measurement.

4.2.3 LIMITATIONS

The most limiting assumption of the whole process is the constant basis spectra assumption, which implies that the workflow does not cover nonlinear effects such as quenching caused by phosphor saturation due to a high radiant flux of the LED [75] or phosphor self absorption. These effects cause a change of the relative phosphor spectrum as a function of angular or spatial dimension.

In the case of phosphor converted white LEDs without additional spectral sources, the influence of such effects can be estimated by obtaining angularly resolved colorimetric measurements[2] as shown in Fig. 4.5. If the measured chromaticities are on a straight line in the CIE xy or CIE $u'v'$ diagram, then the basis spectra can be assumed to be constant because the additive color mixing changes the chromaticity only between the chromaticity coordinate of the blue LED and the chromaticity coordinate of the yellow phosphor. This is for instance the case in Fig. 4.5 (a).

[2] Often strong color variations occur in dark areas. Therefore attention has to be paid during the measurements to avoid a decision which is based on noise.

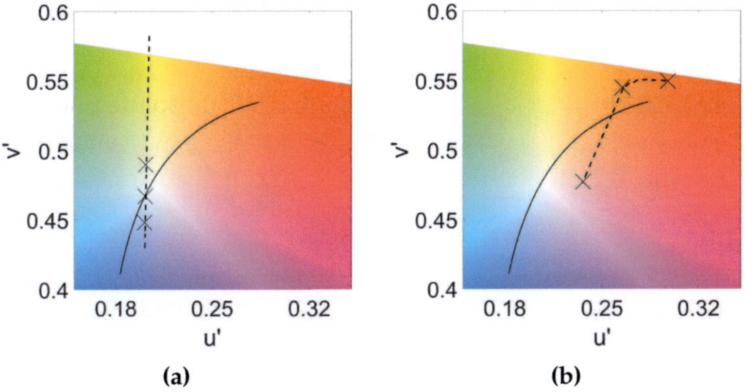

Figure 4.5: Schematic angular color uniformity of two different phosphor converted white LEDs:
(a) Linear shift - 2 basis spectra assumption is justified.
(b) Nonlinear shift - 2 basis spectra assumption is not justified.

If the measured chromaticities of a system with two basis spectra do not lie on a straight line in the CIE xy or CIE $u'v'$ diagram as in Fig. 4.5 (b), the PMBS approach should not be used because there are nonlinear effects. Typically, the nonlinear effects are weak compared to the effect of the linear basis spectra combination. Note that Fig. 4.5 (a) is based on real measurement data but Fig. 4.5 (b) displays only an artificially produced example.

Chromatic effects resulting from a primary optic cannot be modeled with the PMBS approach. Although it is common in near field photometry, it shall be mentioned that all interrelated measurements require the steady state condition with respect to electrical and thermal operation conditions.

4.2.4 WORKFLOW OF PMBS

Based on section 4.2, a general workflow for PMBS can be concluded. The workflow is visualized in Fig. 4.6 and serves as an outline for the next chapters.

The process starts with the determination of the individual semiconductor and phosphor basis spectra using a spectral measurement. The determination of the n_{max} basis spectra is discussed in detail in chapter 5. Subsequently, the goniophotometric measurement with n_{max} different suitable filters are performed. Chapter 6 will focus on a precise definition of the term "suitable" in this context. Then the obtained results in the form of ray files and the spectral sensitivities of the measurement system and the basis spectra are combined to form the spectral reconstruction matrix $M_{S\tau}^{-1}$. Subsequently, the ray files are used to solve the system of linear equations by applying the reconstruction equation Eq. 4.17. While this step is mathematically unproblematic as described above, the practical application on ray files is not trivial because a ray file is a randomized statistically discrete set of vectors. Therefore, chapter 7 will focus on the practical application of the reconstruction equation Eq. 4.17.

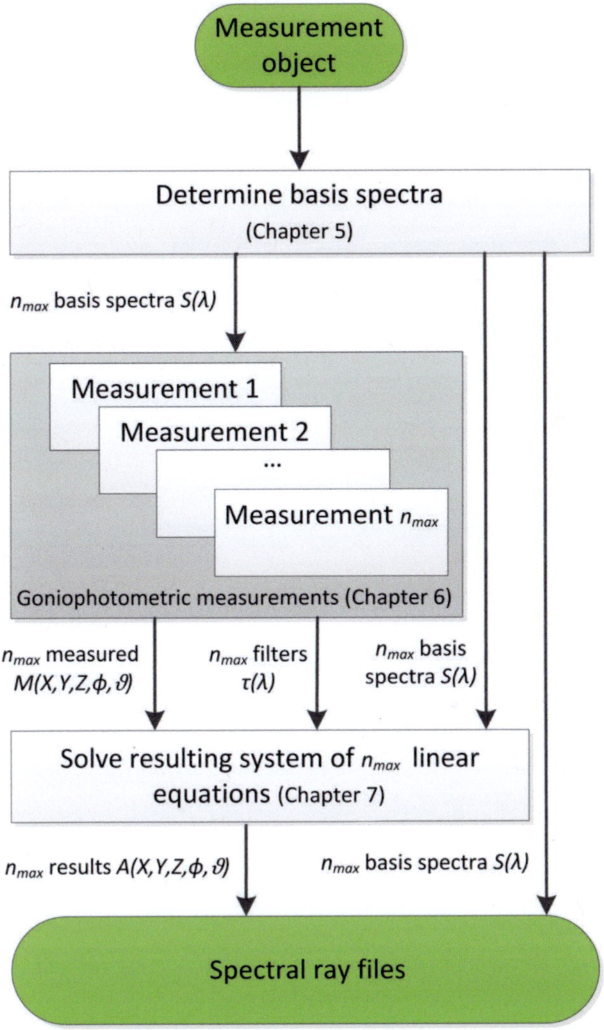

Figure 4.6: Workflow of physically motivated basis spectra (PMBS) approach.

Chapter 5

PMBS: Modeling the Basis Spectra

In a best case scenario, the semiconductor and phosphor basis spectra, which are necessary to set up the reconstruction matrix of the PMBS method, would already be available due to prior knowledge of the system. Then this chapter might be skipped. However, more often than not they are not available. Therefore, this chapter describes an algorithm, which estimates the basis spectra based on predefined phenomenological LED models and one spectral measurement. While this "initial modeling" performs well in most cases, complex spectral systems can be optimized further if an additional assumption and the relations from chapter 4 are used.

5.1 Phenomenological basis spectra of LEDs

Each spectral model aims to describe the spectral power distribution of an LED semiconductor as accurately as possible. The difference between a phenomenological model and a physical model is its foundation. A phenomenological modeled LED spectrum describes the correct spectral power distribution with a solely mathematical distribution. A physically modeled LED spectrum describes the correct spectral power distribution based on physical parameters.

Before the basis spectra estimation process is introduced in the next section, some possible LED basis spectra functions will be discussed.

There are plenty of different phenomenological LED models in the literature [23], [76]–[80]. Reifergerste et. al. summed up and evaluated some of them [81]. Table 5.1 shows the models which were considered in this thesis. Each model has a parameter which represents the peak wavelength λ_C of the LED and an initial amplitude p. In the case of PMBS and according to section 4.2.2 and appendix A.1, the initial amplitude is unimportant as long as $p \in (0, \infty)$. The remaining parameters σ_n describe the width and skewness of each model function. Their number differs between the model functions.

Table 5.1: Potential model functions for LED spectra (extract from [44], [81])

Name	Basis spectrum $S_n(\lambda) = p\times$
Gaussian	$e^{-\left(\frac{\lambda-\lambda_C}{\sigma_1}\right)}$
Second order Lorentzian	$\frac{1}{\left[1+\left(\frac{\lambda-\lambda_C}{\sigma_1}\right)^2\right]^2}$
Logistic power peak	$\left(1+e^{\psi}\right)^{-\frac{\Psi}{\sigma_2}} \left(e^{\psi}\right) \frac{\Psi}{\sigma_2}^{\frac{\Psi}{\sigma_2}}$
Asymmetric logistic peak	$\left(1+e^{-\psi}\right)^{-\Psi} \left(e^{-\psi}\right) \sigma_2^{-\sigma_2} \left(\Psi\right)^{\Psi}$
Pearson VII	$\frac{1}{\left[1+\left(\frac{\lambda-\lambda_C}{\sigma_1}\right)^2\left(\sqrt[\sigma_2]{2}-1\right)\right]^{\sigma_2}}$
Asymmetric double sigmoidal	$\frac{1}{1+e^{-\left(\frac{\lambda-\lambda_C+\sigma_3}{\sigma_1}\right)}}\left(1-\frac{1}{1+e^{-\left(\frac{\lambda-\lambda_C-\sigma_3}{\sigma_2}\right)}}\right)$
Boltzmann based [82]	$S_n(\nu) = p \times \left(e^{\left(-\sigma_1(\nu-\nu_C)\right)} + \sigma_2 e^{\left(\frac{h}{k\sigma_3}(\nu-\nu_C)\right)}\right)^{-1} + \Xi$
with $\psi = \frac{(\lambda-\lambda_C+\sigma_1 \ln\sigma_2)}{\sigma_1}$ and $\Psi = \sigma_2+1$ and $\Xi = \sigma_4 e^{-\left(\frac{\nu-\sigma_5}{\sigma_6}\right)}$	

An important difference exists between symmetric and asymmetric models. In the case of symmetric models, such as for example a Gaussian or a second order Lorentzian distribution, it is assumed that the distribution is symmetric towards the peak wavelength λ_C. This is not true for real LED spectra and therefore the models lack accuracy [81]. But it is advantageous that the models are very simple and that the parameter σ_1 can be interpreted directly as the width of the spectral power distribution. The Pearson VII distribution is also symmetric and in the case of $\sigma_2 = 2$ identical to the second order Lorentzian (σ_1 differs only by a factor).

The asymmetric models are in general more precise in describing the spectral power distribution than the symmetric models, but they are also more complicated. The skewness is described by at least two parameters σ_n. The phenomenological Logistic power peak function and the physically modeled Boltzmann based distribution (which is also the only one which models the spectrum as a function of frequency ν) have the advantage of already having been extended to describe thermal and in the case of the Logistic power peak also current dependencies of the spectra [81], [82]. The Logistic power peak and the Asymmetric logistic peak require only one more parameter than the symmetric models, which is an advantage in the case of numerical optimization techniques.

In contrast, the Boltzmann based model, which is shown in the last row of table 5.1, requires five more parameters than the Logistic power peak. Three of them are needed to fit the Gaussian distribution (the second term), which is added to the Boltzmann based first term to adjust the model to measurement data. In practice the second term is obtained in a second fitting process after the result of the first fit (first term) is subtracted from the measurement data [82]. Therefore, this additional Gaussian slightly reduces the physical background and adds a phenomenological part to that model.

In the case of PMBS, the parameters of the models should be characteristic and assignable to a specific feature of the spectral power distributions. Most spectra in table 5.1 fulfill this condition. This is important because it simplifies the differentiation of the LED spectra from each other and from the phosphor spectrum and enables the definition of starting parameters and constraints for optimization algorithms. Furthermore, it is an advantage for optimization techniques, if the number of model parameters remains small. Note that piecewise defined functions are not considered as LED model spectra as they do not assess specific features to parameters, which makes them inappropriate for the optimization discussed in the next subsections.

5.2 INITIAL MODELING

This section is divided into three subsections. The first describes the modeling process of a single spectrum, the second the slightly more complex case of overlapping LED spectra and the third the most complex case which is LED spectra and a phosphor.

5.2.1 SINGLE LED SPECTRUM

The aim of the initial modeling is to estimate the model parameters $x = \{p, \lambda_C, \sigma_1 ... \sigma_n\}$ for each spectral source. The most straightforward approach is to measure each spectral source in the system individually and fit a parametrized model. An optimal fit is obtained by finding the minimal residual sum of squares RSS, which is defined as

$$RSS = \min_{x \in \mathbb{R}} f(x) \tag{5.1}$$

with

$$f(x) = \sum_{\lambda} (S_{\mathrm{M}}(\lambda) - S(\lambda, x))^2 \tag{5.2}$$

by directly using an optimization algorithm.

Figure 5.1 compares measured LED spectra $S_{\mathrm{M}}(\lambda)$ to fitted LED spectra $S(\lambda, x)$ for the different model functions. The Boltzmann based model is ignored because of its high number of parameters and due to the two-step fitting process. A more detailed explanation for neglecting the Boltzmann based model is given in subsection 5.2.3.

Each colored LED spectrum (red, green and blue) was measured and fitted separately. The figure shows the sum of the measurements and the sum of the models as well as their difference. It can be seen that there is practically no difference between the different LED spectra (red, green and blue). Furthermore, the figure shows that there is no important difference between the different model functions apart from the higher performance of the asymmetric models (compared to the symmetric models) and the low performance of the Gaussian compared to all other model functions.

Table 5.2 summarizes the results for all tested model functions and the individual LEDs. To allow a better comparison of the different models, the RSS is normalized to the RSS of the Gaussian model RSS_{G}. The model performance is better if the RSS is smaller. The additionally provided chromaticity distance $\triangle u'v'$ helps to assess the physiological impact of the fitting error because the residual sum of squares has no physiological meaning. Note that a difference in $\triangle u'v'$ alone is prone to metamerism and can theoretically give a value $\triangle u'v' = 0$ although the fit shows clear differences. The analysis of both figures of merit (RSS and $\triangle u'v'$) leads to the conclusion that apart from the Gaussian, all tested model spectra might be used to model individually measured LED spectra (at least those without a phosphor).

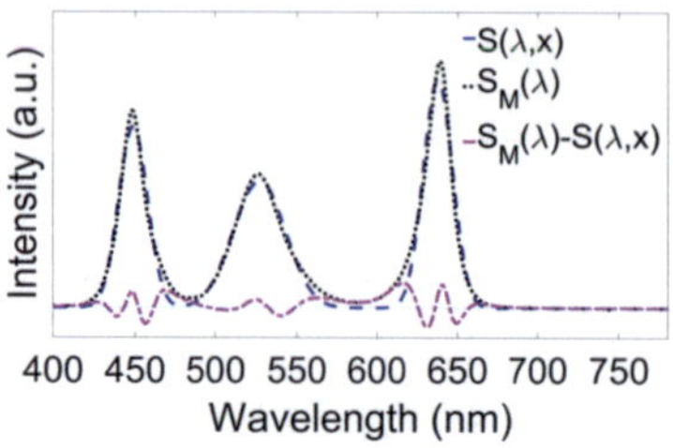

(a) Gaussian

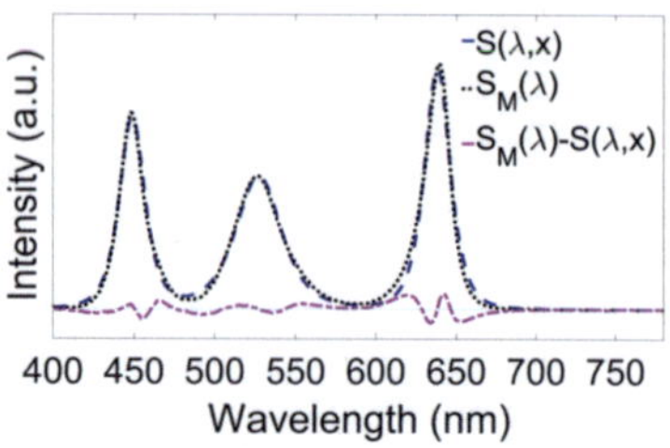

(b) Second order Lorentzian

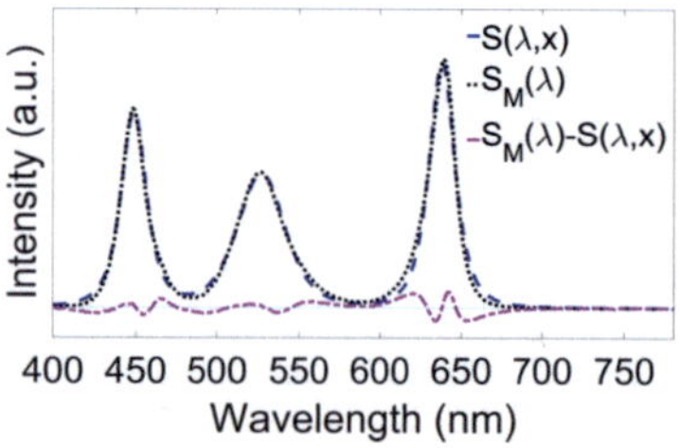

(c) Pearson VII

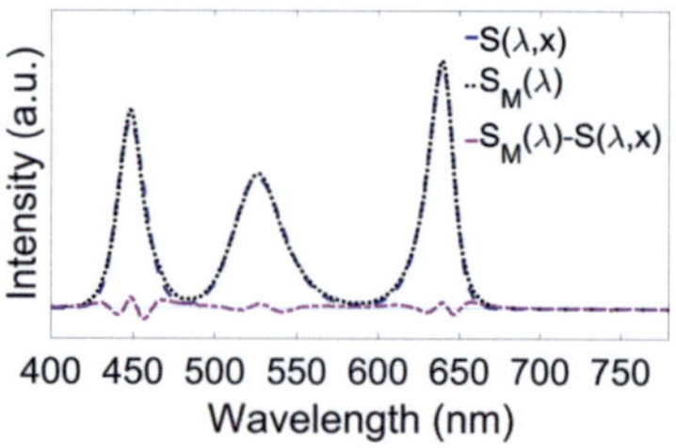

(d) Asymmetric logistic peak

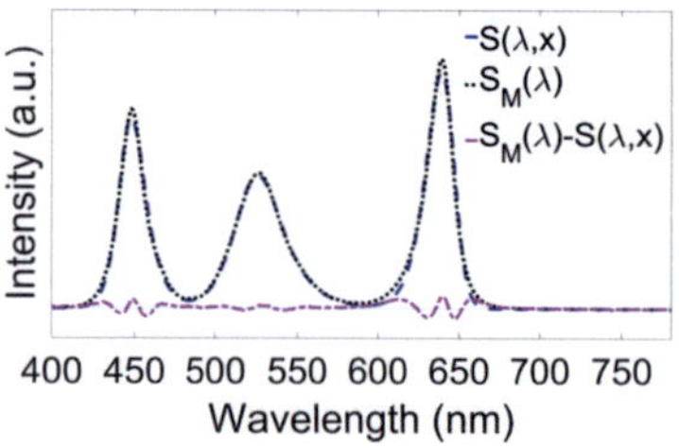

(e) Logistic power peak

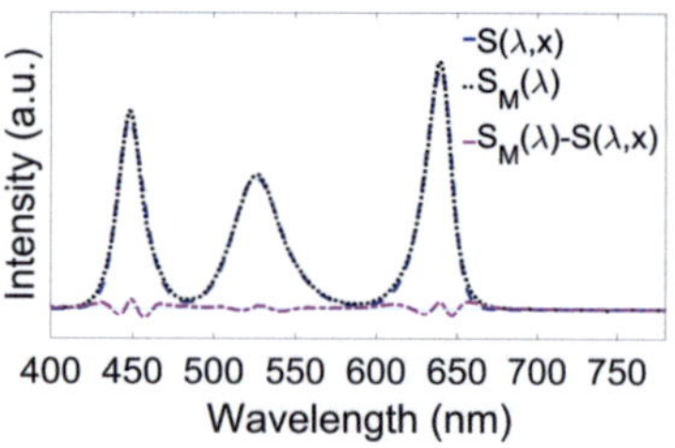

(f) Asymmetric double sigmoidal

Figure 5.1: Comparison of measured spectra $S_{\mathrm{M}}(\lambda)$ and fitted LED spectral models $S(\lambda, x)$ for different model functions.

Table 5.2: Goodness of fit of individually measured LED spectra

Name	Blue		Green		Red		Sum
Figure of merit	$\frac{RSS}{RSS_G}$	$\triangle u'v'$	$\frac{RSS}{RSS_G}$	$\triangle u'v'$	$\frac{RSS}{RSS_G}$	$\triangle u'v'$	$\frac{RSS}{RSS_G}$
Gaussian	1	0.023	1	0.011	1	0.022	1
Second order Lorentzian	0.27	0.015	0.53	0.006	0.49	0.006	0.37
Pearson VII	0.25	0.010	0.49	0.005	0.48	<0.001	0.37
Logistic power peak	0.19	0.016	0.05	0.001	0.27	0.011	0.21
Asymmetric logistic peak	0.35	0.019	0.14	0.004	0.07	0.003	0.16
Asymmetric double sigmoidal	0.23	0.017	0.07	0.002	0.10	0.005	0.13

However, as discussed at the beginning of section 3.4, an individual measurement can be problematic in practical applications and may even affect the shape of the obtained spectral power distribution. Therefore an estimation has to be possible if only one measured spectrum is available.

5.2.2 Overlapping LED spectra

Due to the spectral overlap the modeling gets more complicated if all spectra are estimated from one measured spectrum, which contains the sum of all basis spectra. Besides the measurement spectrum $S_M(\lambda)$, the proposed algorithm requires prior knowledge of the system, which is the number of different LED spectra $n_{\max}$. This knowledge is typically directly available. The optimization problem is the estimation of $n_{\max}$ different parameter sets $x_i = \{p_i, \lambda_{i,C}, \sigma_{i,1}...\sigma_{i,n}\}$ such that $x = \{x_1...x_{n_{\max}}\}$ using the already defined Eq. 5.1 and Eq. 5.2 from the last subsection.

Table 5.3 shows the results of the optimization based on the sum of the individually measured spectra. The RSS is again normalized by the RSS_G of the individual measurements to allow a comparison between the different models as well as to the models obtained from the individual measurements. The last column shows that the RSS of the sum in table 5.2 remains or improves, which is a direct result of the optimization. In the case of the individual "true" basis spectra, the result differs.

Table 5.3: Goodness of fit to sum of measured LED spectra

Name	Blue		Green		Red		Sum
Figure of merit	$\frac{RSS}{RSS_G}$	$\Delta u'v'$	$\frac{RSS}{RSS_G}$	$\Delta u'v'$	$\frac{RSS}{RSS_G}$	$\Delta u'v'$	$\frac{RSS}{RSS_G}$
Gaussian	1	0.023	1.04	0.010	1	0.022	0.99
Second order Lorentzian	0.30	0.016	0.56	0.006	0.49	0.006	0.37
Pearson VII	0.33	0.018	0.69	0.008	0.48	0.001	0.36
Logistic power peak	0.20	0.016	0.07	<0.001	0.27	0.011	0.21
Asymmetric logistic peak	0.35	0.019	0.19	0.004	0.07	0.002	0.16
Asymmetric double sigmoidal	0.23	0.017	0.08	0.001	0.10	0.005	0.13

The most interesting model in this comparison is Pearson VII. The *RSS* of the blue and green LED models worsens, while the *RSS* of the red LED stays almost constant. The blue and green LED are more affected because they have a stronger overlap than the green and red LED model, which can be seen in Fig. 5.1. It is also important to note that the individual Pearson VII models deteriorate more than the individual Lorentzian models (while the *RSS* of the sum stays better). All other models show only small differences between the individual fits (table 5.2) and the fits from the sum of the individual spectra (table 5.3).

As long as the spectral overlap of the individual spectral sources is small, the optimization of the sum will provide results as good as those from an individual fit. But the result may get worse as consequence of a spectral overlap as in the case of Pearson VII. If the spectral overlap becomes larger, it is helpful to use models with less optimization parameters and easily definable constraints as for instance the Lorentzian function. Nevertheless, the comparison shows that in simple cases without a phosphor the basis spectra obtained are good enough to proceed with PMBS directly.

5.2.3 SPECTRAL SOURCES WITH PHOSPHOR

In case of the presence of a phosphor the optimization becomes more complicated for two reasons:

- The spectral overlap between a phosphor and a LED spectrum can be very large (as in the case of a green or red LED and a yellow phosphor, such as in RGBW systems).
- There is no general phenomenological parameterized model function, which can describe the spectral power distribution of a phosphor.

Again the following approach requires prior knowledge of the system, which is the number of basis spectra n_{max} and the occurrence of the phosphor as well as the measured spectrum $S_{\text{M}}(\lambda)$. Due to the occurrence of the phosphor the number of LED basis spectra is reduced to $n_{\text{max}} - 1$. Consequently, the following approach assumes that the phosphor spectrum can be described by one distribution. The modeling procedure visualized in Fig. 5.2 consists of a four step process:

Step 1 is similar to the optimization without a phosphor. The measured spectrum $S_{\text{M}}(\lambda)$ is used to estimate n_{max} LED basis spectra by using Eq. 5.1 and Eq. 5.2, although it is known that there are only $n_{\text{max}} - 1$ LED spectra. Then the obtained model parameters are used to identify the model spectrum having the least probability of being an LED spectrum. Hence a characteristic model parameter is required. In the case of a Lorentzian this would be σ_1 as it describes the width of the model spectrum. The broadest spectrum is assumed to be the phosphor.

Step 2 directly follows the decision which assigns the individual models to the LED and phosphor. The remainder phosphor spectrum $S_{\text{R,P}}(\lambda)$, which is defined as

$$S_{\text{R,P}}(\lambda) = S_{\text{M}}(\lambda) - \sum_{n=1}^{n_{\text{max}}-1} S_n(\lambda, x), \tag{5.3}$$

serves as the basis of the phosphor. A smoothed spline is used to describe the phosphor. The smooth spline is used to represent the general distribution of the phosphor because it models the only known general attributes, which are smoothness and continuity, with the highest precision. The smoothing parameter p_{smooth} is an optimization variable. However, as $p_{\text{smooth}} = 1$ would equal a cubic spline, it needs a user defined upper constraint to maintain the smoothness. The result is a first estimate of the phosphor spectrum $S_{\text{P}}(\lambda)$.

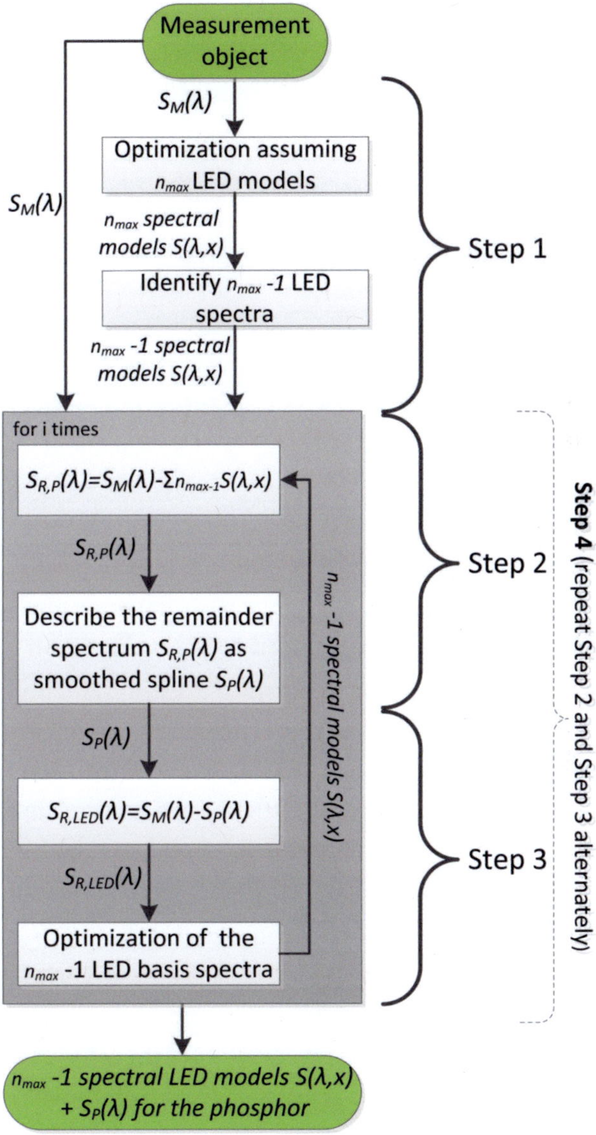

Figure 5.2: Estimation of basis spectra if a phosphor spectrum occurs.

Step 3 is similar to step 2 but focuses on the LED instead of the phosphor. The phosphor spectrum $S_P(\lambda)$ is used to recalculate the $n_{max} - 1$ LED spectra, which are based on the remainder spectrum $S_{R,LED}(\lambda)$ defined as

$$S_{R,LED}(\lambda) = S_M(\lambda) - S_P(\lambda). \tag{5.4}$$

$S_{R,LED}(\lambda)$ and the former $n_{max} - 1$ LED spectra are then used in the optimization Eq. 5.1 and Eq. 5.2. The result is a set of $n_{max} - 1$ improved LED models $S_n(\lambda, x)$.

Step 4 is the iterative improvement of the model spectra by alternately repeating step 2 and step 3. The results of step 3 are used to revise $S_{R,P}(\lambda)$ and the spline $S_P(\lambda)$ and subsequently, the results of step 2 can again be used to revise the LED basis spectra. Step 4 can be repeated until it converges or until a predefined number is reached as indicated by Fig. 5.2. Finally, the n_{max} basis spectra are given by the $n_{max} - 1$ LED basis spectra and $S_P(\lambda) = S_{n_{max}}(\lambda)$.

The approach is similar to obtaining the Boltzmann based model described in [82] which models a Gaussian distribution that is also based on a remainder spectrum in a second optimization. This is also the reason why the Boltzmann based model is not used in this thesis apart from its high number of model parameters. Modeling the phosphor would lead to a more complex 3-step process and the additional Gaussian might disturb the phosphor spectrum.

Figure 5.3 visualizes each step with the example of a phosphor converted white LED with an additional red LED using the Logistic power peak model. In the first step, the sum of three Logistic power peaks is fitted to the measured spectrum $S_M(\lambda)$ as shown in Fig. 5.3 (a). Then it is decided which LED basis spectrum has the least probability of being an LED basis spectrum, which is $S_2(\lambda, x)$.

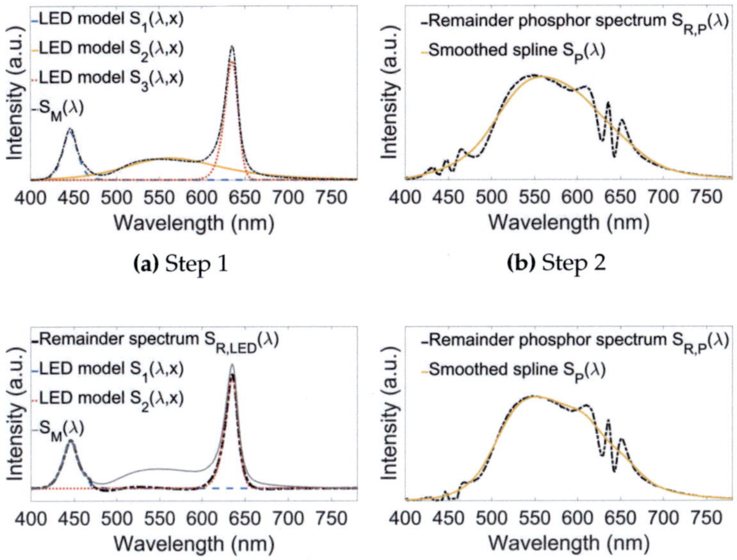

(a) Step 1 **(b)** Step 2

(c) Step 3 **(d)** Step 4

Figure 5.3: Example of basis spectra estimation if a phosphor spectrum occurs:
Step 1: First estimation according to Eq. 5.1 and Eq. 5.2 based on the measured spectrum $S_M(\lambda)$. As the LED model spectrum $S_2(\lambda, x)$ is the broadest, it is assumed to be the that of the phosphor.
Step 2: Calculate remainder phosphor spectrum $S_{R,P}(\lambda)$ as the difference between the measured spectrum $S_M(\lambda)$ and $S_1(\lambda, x) + S_3(\lambda, x)$ and fit the smoothed spline $S_P(\lambda)$.
Step 3: Re-estimation according to Eq. 5.1 and Eq. 5.2 based on the remainder spectrum $S_{R,LED}(\lambda)$ of the LEDs.
Step 4 (first part)**:** Recalculate remainder phosphor spectrum $S_{R,P}(\lambda)$ and smoothed spline $S_P(\lambda)$ and proceed as shown in Fig. 5.2.

The characteristic inhomogeneities of the phosphor remainder spectrum $S_{R,P}(\lambda)$ around the peak wavelength of the LED spectra are caused by the difference between the LED model spectra and the measured spectrum. They are similar to the differences observed in Fig. 5.1. The smoothed spline $S_P(\lambda)$ with the smoothing parameter p_{smooth} avoids the inhomogeneities.

Figure 5.3 (c) shows that the blue LED spectrum changes slightly if $S_P(\lambda)$ is considered. Consequently, $S_P(\lambda)$ also changes slightly around the blue LED spectrum compared to Fig. 5.3 (b) in the next step.

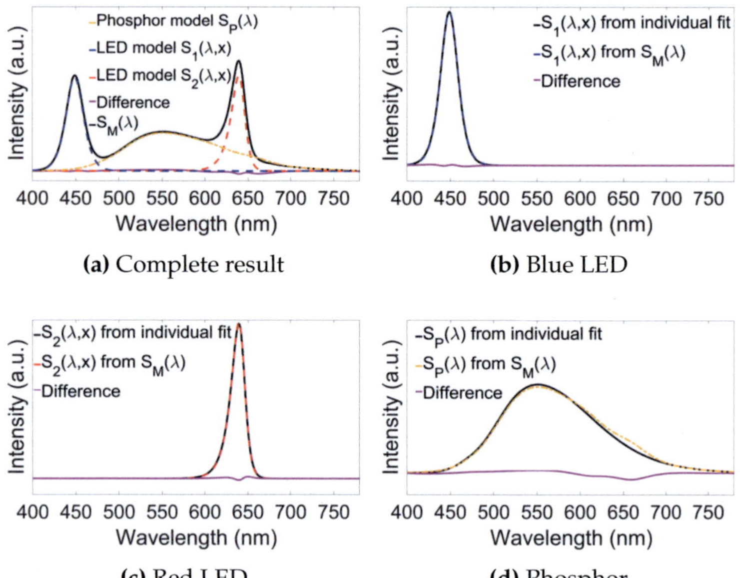

(a) Complete result

(b) Blue LED

(c) Red LED

(d) Phosphor

Figure 5.4: Results of basis spectra estimation of white (blue/phosphor) and red LEDs compared to the individually fitted basis spectra.

Figure 5.4 shows results obtained from the combination of red LED, blue LED and phosphor (RBY). The figure displays the fitted models from the individual measurements compared to those estimated from the sum as well as the corresponding difference. In the figure, the Asymmetric logistic peak was used.

The largest difference can be observed between the phosphor spectrum in the spectral regime which overlaps to the red LED. But the absolute difference is small for all three basis spectra. This is also supported by the numeric comparison of the absolute chromaticity distances $\triangle u'v'$ shown in table 5.4. Note that the chromaticity distance of the blue LED and the phosphor refer to a fit which was obtained by using only the measured spectrum of the white LED. The chromaticity distance of the red and white LED refers to the distance to the individually measured spectra.

Figure 5.5 shows results obtained in the case of the combination of green LED, blue LED and phosphor (GBY). The spectral overlap between the green LED spectrum and the phosphor spectrum is larger than in the previous RBY case. The individual peaks of the phosphor and the green LED even merge to one as shown in Fig. 5.5 (a). This example is therefore more critical. Nevertheless, the obtained basis spectra of the blue and green LED in sub Fig. 5.5 (b) and (c) match those of the individual measurements.

Only the phosphor spectrum in sub Fig. 5.5 (d) shows a clear displacement and a large chromaticity distance $\triangle u'v'$ (see also table 5.4). The phosphor displacement is mainly caused by a wrong estimation of the amplitude ratios, which shifts the phosphor peak towards higher wavelengths and changes it especially in the regions of stronger overlap. The RSS still remains high as the sum matches the measurement. The problem is the same as that which caused the deterioration of the Pearson VII model described in subsection 5.2.2. However, the effect is stronger because the spectral overlap of the green LED and the phosphor is much stronger.

Table 5.4: Comparison of fitted model spectra to individually measured spectra in $\triangle u'v'$

<table>
<tr><th rowspan="3">Name</th><th colspan="3">RBY</th><th colspan="3">GBY</th></tr>
<tr><th rowspan="2">R</th><th colspan="2">W</th><th rowspan="2">G</th><th colspan="2">W</th></tr>
<tr><th>B</th><th>Y</th><th>B</th><th>Y</th></tr>
<tr><td rowspan="2">Second order Lorentzian</td><td rowspan="2">0.003</td><td colspan="2">0.004</td><td rowspan="2">0.012</td><td colspan="2">0.036</td></tr>
<tr><td><0.001</td><td>0.004</td><td>0.001</td><td>0.044</td></tr>
<tr><td rowspan="2">Logistic power peak</td><td rowspan="2">0.012</td><td colspan="2">0.010</td><td rowspan="2">0.004</td><td colspan="2">0.026</td></tr>
<tr><td>0.005</td><td>0.012</td><td>0.005</td><td>0.033</td></tr>
<tr><td rowspan="2">Asymmetric logistic peak</td><td rowspan="2">0.002</td><td colspan="2">0.004</td><td rowspan="2">0.002</td><td colspan="2">0.018</td></tr>
<tr><td>0.001</td><td>0.005</td><td>0.004</td><td>0.022</td></tr>
</table>

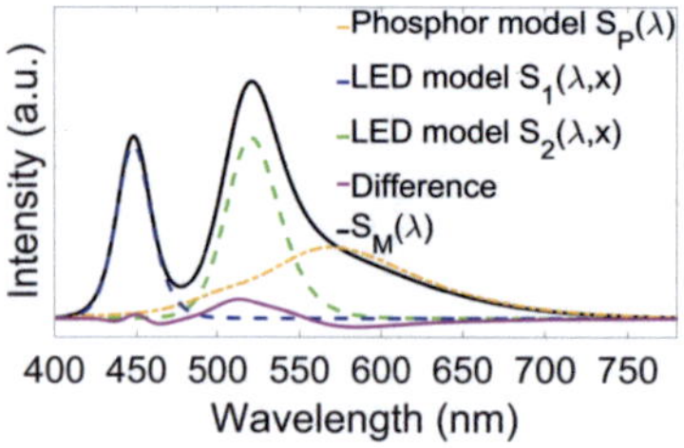

(a) Complete result

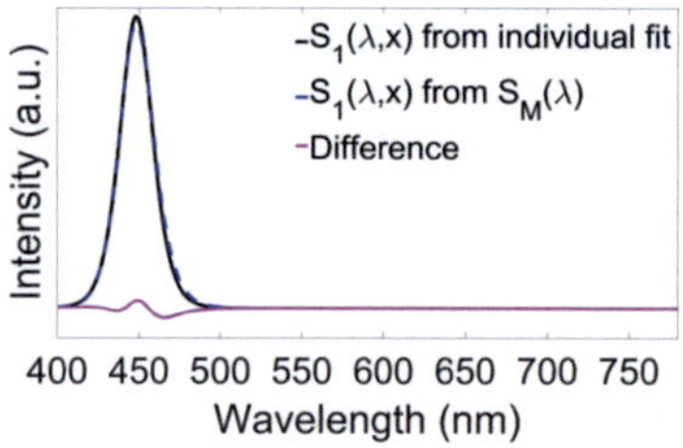

(b) Blue LED

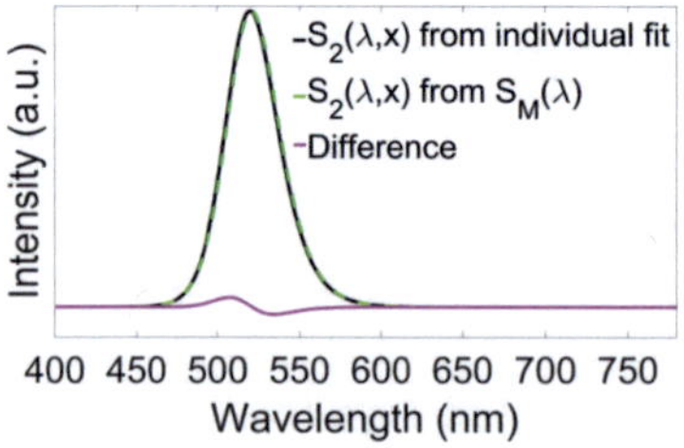

(c) Green LED

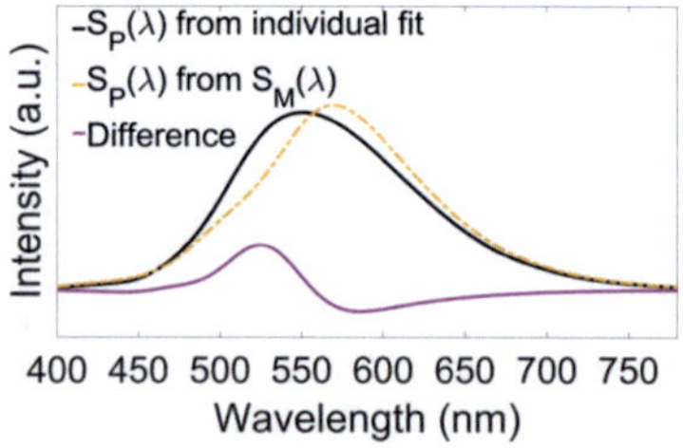

(d) Phosphor

Figure 5.5: Results of basis spectra estimation of white (blue/phosphor) and green LEDs compared to the individually fitted basis spectra.

To sum up: Although the estimation provides useful results in most cases, such as phosphor converted white LED or a combination of phosphor converted white LED and red LED, the basis spectra obtained differ stronger in the case of a green LED and a phosphor.

The limitations occur due to the limited physical background of the chosen figure of merit *RSS* in Eq. 5.1 and Eq. 5.2, which only ensures a correct sum. Furthermore, the approach cannot be used if the phosphor description requires more than one phosphor modeling function, as for instance in the case of an inhomogeneous combination of different phosphor layers or the combination of warm white and cold white LED. However, by using additional prior knowledge the estimation can be optimized using the PMBS approach itself.

5.3 BASIS SPECTRA VALIDATION AND OPTIMIZATION

Until this section, chapter 5 only serves as a basis for the other chapters of this thesis. The basis spectra are estimated by optimizing the residual sum of squares from the sum of all basis spectra and a measurement. But the optimization is solely mathematically motivated and lacks a physical background. As a result, the individual basis spectra differ as for instance in the case of GBY. Furthermore, it is not simple to choose a set of spectra obtained from different optimizations (for instance different LED model functions), if the individual spectra cannot be measured (if they could be measured there would be no need to estimate them in the first place). As the reconstruction matrix of PMBS requires the correct basis spectra, the approach would be flawed were incorrect basis spectra to be used. Therefore, it is necessary to validate the basis spectra. This section describes possibilities of validating the basis spectra based on measurements in the goniometric setup. They allow the choice of the most suitable set of basis spectra.

Finally, these validations may be used as additional merit functions to optimize basis spectra, which were for instance obtained during the solely mathematical initial modeling described in the last section.

5.3.1 SPATIAL SEPARATION

This validation incorporates an additional physical attribute of LED systems, which is the spatial separation of colored LEDs (often also the phosphor as part of only the white/blue LED). Most often, the different LED colors lie next to each other. A green LED is green, a red LED is red and a white LED is white (or rather blue and yellow). The green, red and white LEDs do not overlap in the 3-dimensional spatial dimension. Although there might be a remote phosphor, which overlaps with all LEDs, the single colored chips are both spatially separated and highly localized in the three dimensional spatial dimension. Those known attributes can be used.

A set of single pseudo-radiance images (or pseudo-irradiances based on measured ray files) $M_n(X,Y)$ is used to reconstruct the single radiance images (or irradiances) $A_n(X,Y)$ of the individual basis spectra using the mathematical relations described in chapter 4 by applying Eq. 4.17. The reconstructed $A_n(X,Y)$ can be used to quantify the "spatial separation" of the individual spectral sources. Often a 2-dimensional spatial separation in the technical LED plane is sufficient. A high spatial separation and localization consists of a few high values and as many zeros or dark areas as possible (the red LED does not lie on a white LED). To quantify the term "as many zeros or dark areas as possible", the figure of merit $P_{\mathrm{sum},n}$ has been developed. The mathematical details and an artificial example are provided in appendix A.2. $P_{\mathrm{sum},n}$ becomes small if there are few high peaks and many values near zero. If there are artifacts (for instance a green LED spatially on top of a white LED) beside the high peaks, the values corresponding to the artifacts result in a higher $P_{\mathrm{sum},n}$. As there are $n_{\max}$ $P_{\mathrm{sum},n}$ (one for each basis spectrum) the final figure of merit is defined by

$$P_{\mathrm{sum}} = \sum_{n=1}^{n_{\max}} P_{\mathrm{sum},n}. \tag{5.5}$$

A lower P_{sum} indicates a better spatial separation and therefore a better set of basis spectra.

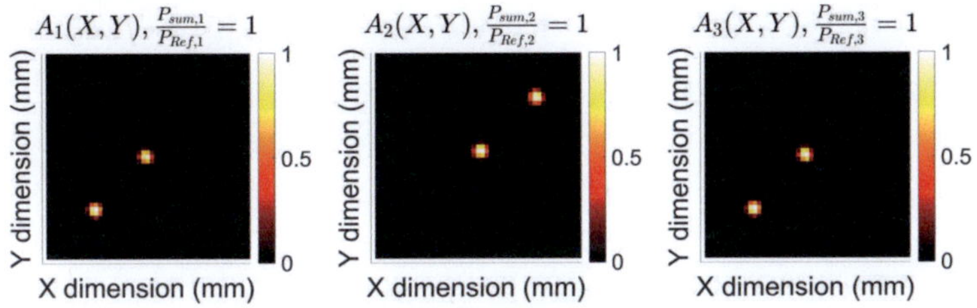

(a) Basis spectra from individual measurements

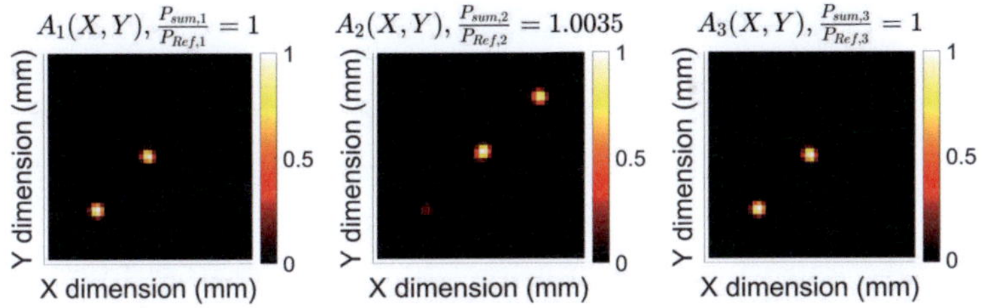

(b) Basis spectra from $S_{\mathrm{M}}(\lambda)$

Figure 5.6: Reconstructed amplitude images $A_n(X,Y)$ using the basis spectra displayed in Fig. 5.5 (from left to right, Blue LED, Green LED, Phosphor).

Figure 5.6 (a) shows the reconstructed normalized $A_n(X,Y)$ for the case of the individually measured models for the GBY LED and Fig. 5.6 (b) the results for the estimated basis spectra using $S_{\mathrm{M}}(\lambda)$ (the spectra are those displayed in Fig. 5.5). They are based on irradiances $M_n(X,Y)$ obtained from measured ray files. The given values of the spatial separation $P_{\mathrm{sum},n}$ are normalized to those of the individual measurements as reference values to allow a better comparison. It can be seen that the measured object contains three spatial regions: the bottom left, the middle and the top right. While the middle region contains white (blue/phosphor) and green LED radiation, the bottom left contains only white (blue and phosphor) radiation and the top right only green LED radiation[1].

As expected, the individual measurements from sub Fig. 5.6 (a) outperform the mathematical deduced spectra from the initial modeling from sub Fig. 5.6 (b). In particular the reconstructed irradiance $A_2(X,Y)$ (the green LED) shows an artifact in the bottom left region in sub Fig. 5.6 (b), which also leads to the higher $P_{\mathrm{sum},2}$ compared to sub figure (a). Note that the spatial resolution of the irradiances should be low to avoid influence of noise caused by high resolutions of the $A_n(X,Y)$.

5.3.2 PUNCTUAL SPECTRAL MEASUREMENTS

Another option to validate the basis spectra would be to perform additional spectral measurements at different angular positions (ϕ, ϑ) before or after the ILMD-based measurements. The ray files can be used to reconstruct the spectra using the PMBS method as described in chapter 4 using Eq. 4.12 and Eq. 4.17 at the measured angular positions (ϕ, ϑ). Then the reconstructed spectra can be compared to the spec-

[1] Of course the two LED types green (5.6 (a), middle) and white (5.6 (a) right and left) are also spatially separated in the middle region. The irradiance images merge only because the plane of projection is above the LED plane. To see the separation clearly, it would be necessary to display the images directly in the LED plane. This is not done in Fig. 5.6 because each LED is as large as one pixel due to the large distance of the two outer regions compared to the small distance of the LEDs within the middle region. On closer inspection one can notice a slight spatial shift between $A_2(X,Y)$ and the other reconstructed images.

tral measurements at these positions (ϕ, ϑ) in terms of chromaticity distance $\triangle u'v'$ or the residual sum of squares RSS. As described in section 4.2.2 the process may take into consideration that the spectral measurement occurred in the near field as the measurement distance and the diameter of the sensor of the spectroradiometer are known during the ray tracing. The ray tracing is necessary to compute the measurement amplitudes as input for PMBS.

However, this method should only be used if the ray files or rather the light intensity distribution at the spectral measurement position does not lie within a high gradient because then the uncertainty of the measurement amplitudes obtained is too large. Furthermore, a comparison to noisy spectral data (for instance measured at the dark areas) should be avoided.

5.3.3 COLOR MIXING LINE

In the case of phosphor-converted white LEDs there is a similar but slightly advanced option compared to the punctual spectral measurements. As described in section 2.2 and subsection 4.2.3 the color shift always lies on a line in the CIE $u'v'$ diagram if nonlinear effects such as self absorption or saturation can be neglected. As PMBS also assumes that the final spectrum consists of a linear combination of the basis spectra (see Eq. 4.12) the reconstructed spectra also lie on a line between the chromaticity coordinates of the associated basis spectra. This means that the alignment of the theoretical mixing line and some measured points also gives a hint of the precision of the basis spectra.

Nevertheless, due to metamerism there is a theoretical chance that incorrect basis spectra could still lead to a good alignment. Therefore, the punctual spectral measurements required to draw the true mixing line should also be used to reconstruct the spectrum at the measurement positions as described above. The color mixing line is a simple and intuitive representation but also limited to cases with only two basis spectra and therefore mainly to phosphor converted white LED. The restrictions regarding the quality of the spectral measurements are identical to those of the punctual spectral measurements.

5.3.4 OPTIMIZATION OF BASIS SPECTRA

In the last three subsections additional figures of merit for the purpose of basis spectra validation have been developed. A logical consequence of their definitions is to use them directly within an optimization approach to further optimize the basis spectra. The spatial separation P_{sum} in particular can be minimized directly because the reconstruction of the $A_n(X,Y)$ depends on the reconstruction matrix $M_{S\tau}^{-1}$, which is a function of the parameters of the basis spectra[2]. As the spatial separation basically optimizes the reconstruction matrix $M_{S\tau}^{-1}$ it focuses on integrated spectral values but not on a spectrally resolved distribution. Therefore the RSS or the punctual spectral measurements are used as a constraint during the optimization to avoid a solution which would only optimize P_{sum} without considering a measured spectral distribution. In this thesis, the constraint was defined as a tolerated maximum factor for the deterioration the RSS. As the results from the initial modeling provide reasonable starting points, local optimization techniques were used.

[2] The relative weighting should be based on the camera fluxes if this method is used because otherwise the optimization is influenced by the deduction of the relative weighting factors, which also rely on the basis spectra in the second method.

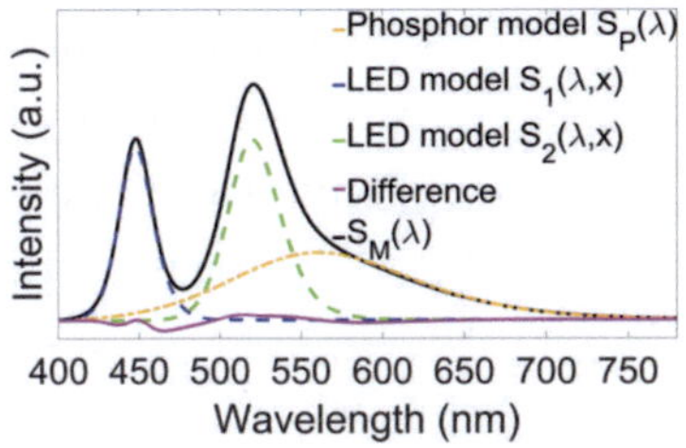

(a) Complete result

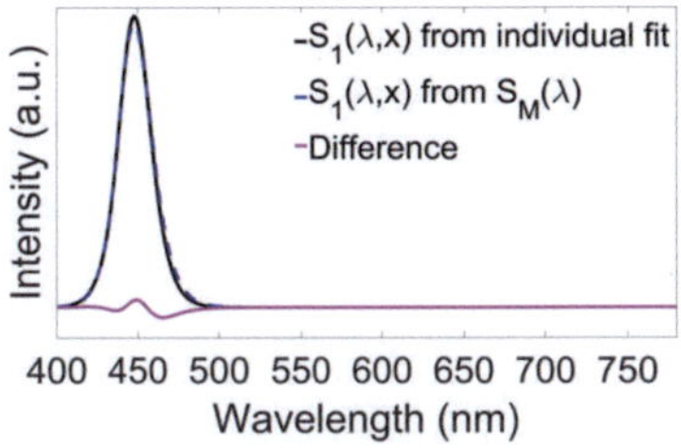

(b) Blue LED

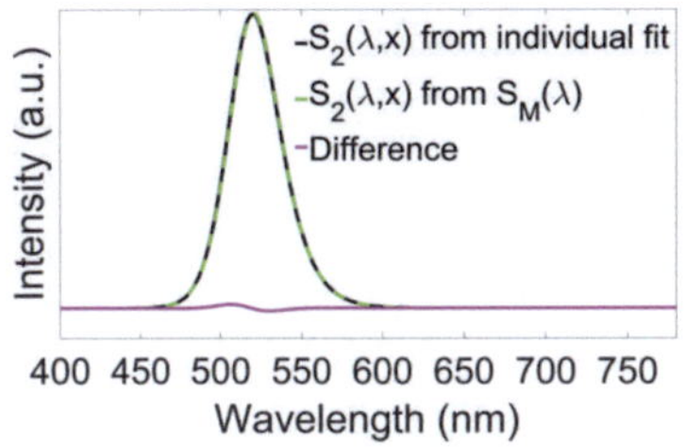

(c) Green LED

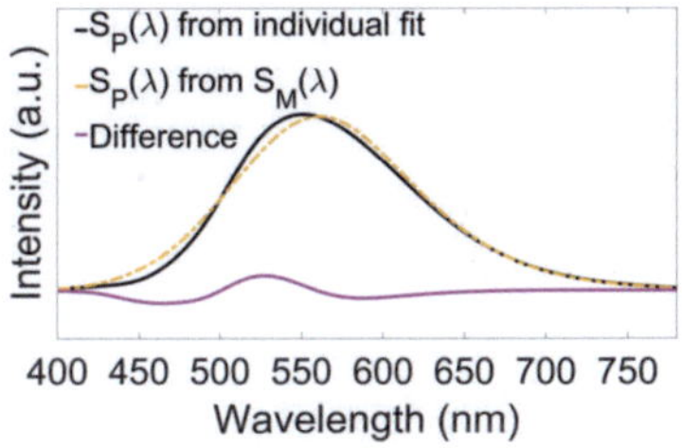

(d) Phosphor

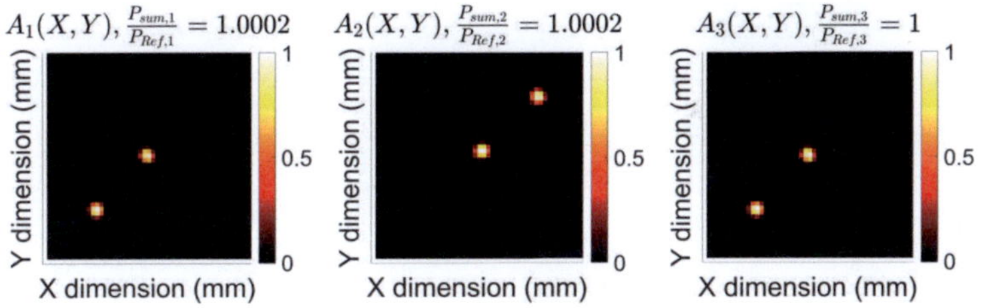

(e) Spatial separation

Figure 5.7: Results of basis spectra estimation using the spatial separation P_{sum} of a combination of white (blue/phosphor) and green LEDs.

Figure 5.7 shows the optimized basis spectra and the spatial separation $P_{\text{sum},n}$ (again normalized to those of the individual measurements as reference) as well as the resulting reconstructed $A_n(X,Y)$ of the GBY Asymmetric logistic peak model. The displacement of the phosphor was improved using this optimization approach. However, as the spatial separation tends to optimize only spatially separated LEDs, the separation of the blue LED and the phosphor may deteriorate.

Table 5.5 shows the results obtained using the different model functions optimized with spatial separation compared to the models obtained from an individual measurement. While the separated green LED performs as well as if it had been based on an individual fit (see also table 5.2), the result from the blue LED deteriorates. However, due to the significant improvement of the phosphor as part of the white LED, the overall results are always better. A more in-depth discussion of the basis spectra optimization is presented in chapter 9 and subsection 10.2.

Table 5.5: Model spectra after (before) optimization using spatial separation compared to individually measured spectra in $\triangle u'v'$

Name	GBY		
	G	W	
		B	Y
Second order Lorentzian	0.003 (0.012)	0.005 (0.035)	
		0.007 (0.001)	0.015 (0.044)
Logistic power peak	0.002 (0.004)	0.009 (0.026)	
		0.003 (0.005)	0.028 (0.033)
Asymmetric logistic peak	0.003 (0.002)	0.006 (0.018)	
		0.004 (0.004)	0.009 (0.022)

CHAPTER 6

PMBS: DETERMINATION OF OPTICAL FILTERS

This chapter shows that the uncertainty of the ILMD-based measurements influences the reconstruction equation of PMBS. It focuses on the selection of the ideal optical filter combination for any arbitrary spectrum of a given filter set. This is done by considering measurement uncertainties of two different filter technologies - glass edge absorption filters and interference filters - with Monte Carlo simulations. A RGBW LED reconstruction with different filters is used to verify the simulations and to underline the importance of the filter selection. Furthermore, a fast preselection technique based on factorial design is introduced to overcome potential high simulation times of Monte Carlo simulations[1].

6.1 MODELING THE UNCERTAINTY OF THE SPECTRAL RECONSTRUCTION

Similarly to the $n_{\max}$ basis spectra, the different $n_{\max}$ effective spectral sensitivities $s_{\tau,\text{eff},n}(\lambda)$ are required to set up the reconstruction matrix $M_{S\tau}$, which is necessary to apply PMBS. According to Eq. 4.13, the effective spectral sensitivity $s_{\tau,\text{eff},n}(\lambda)$ of the ILMD-based goniophotometric measurement depends on the filter transmission $\tau_n(\lambda)$, the neutral density filter transmission $\tau_{\text{ND},n}(\lambda)$ and the remaining system

[1] *Parts of this chapter have already been pubslished in [83].*

sensitivity $s_{\mathrm{Sys}}(\lambda)$. In contrast to the basis spectra the determination of $s_{\tau,\mathrm{eff},n}(\lambda)$ is straightforward. Each filter transmission $\tau_n(\lambda)$ and $\tau_{\mathrm{ND},n}(\lambda)$ as well as the sensitivity $s_{\mathrm{Sys}}(\lambda)$ can be measured individually, as for instance described in [84]–[87].

The effective spectral sensitivities $s_{\tau,\mathrm{eff},n}(\lambda)$ also influence the measurement values $M_n(X,Y,Z,\phi,\vartheta)$ which are required to reconstruct the radiometric amplitudes $A_n(X,Y,Z,\phi,\vartheta)$. Then, according to chapter 4 and Eq. 4.16, the only requirement of the effective spectral sensitivities is fulfilled if $s_{\tau,\mathrm{eff},1}(\lambda) \neq \ldots \neq s_{\tau,\mathrm{eff},n_{\mathrm{max}}}(\lambda)$. This means that at least in theory any arbitrary filter combination is sufficient for a spectral reconstruction with PMBS.

An example is shown in Fig. 6.1. The shown RGBW LED system consists of four basis spectra and, therefore, PMBS requires four different ILMD-based filter measurements. According to the ideal reconstruction Eq. 4.16, the four basis spectra can be reconstructed with, for instance, the filter combinations A-D. However, at a closer look no one would consider to choose filter combination B or C during the application of PMBS in this example because they do not provide an accurate measurement of the blue part of the spectrum.

During the application of PMBS the ideal reconstruction Eq. 4.16 has to be extended to the real reconstruction Eq. 6.1. This equation considers that the reconstructed $A_n(X,Y,Z,\phi,\vartheta)$ differ by $\Delta A_n(X,Y,Z,\phi,\vartheta)$ due to the deviations $\eta_n(X,Y,Z,\phi,\vartheta)$ of the measurements $M_n(X,Y,Z,\phi,\vartheta)$. Note that in the context of PMBS all values in Eq. 6.1 are functions of the spatial and angular dimensions. However, as this dependence is not important in this chapter, the equations are simplified by using only single values.

$$\begin{bmatrix} M_{\mathrm{A},1} + \eta_1 \\ \vdots \\ M_{\mathrm{A},n_{\mathrm{max}}} + \eta_{n_{\mathrm{max}}} \end{bmatrix} = M_{\mathrm{S}\tau} \cdot \begin{bmatrix} A_1 + \Delta A_1 \\ \vdots \\ A_{n_{\mathrm{max}}} + \Delta A_{n_{\mathrm{max}}} \end{bmatrix} \tag{6.1}$$

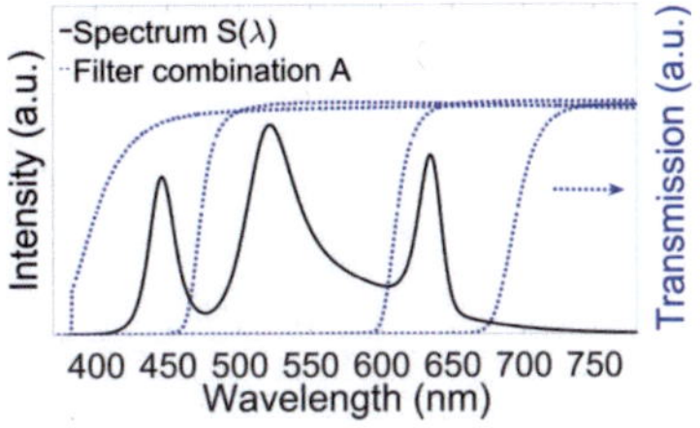

(a) Filter combination A

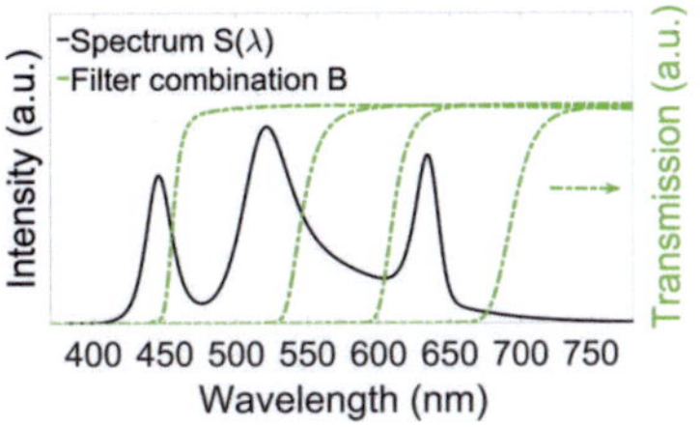

(b) Filter combination B

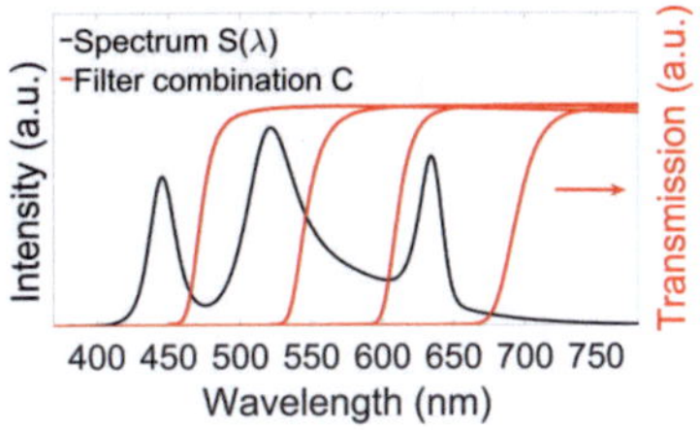

(c) Filter combination C

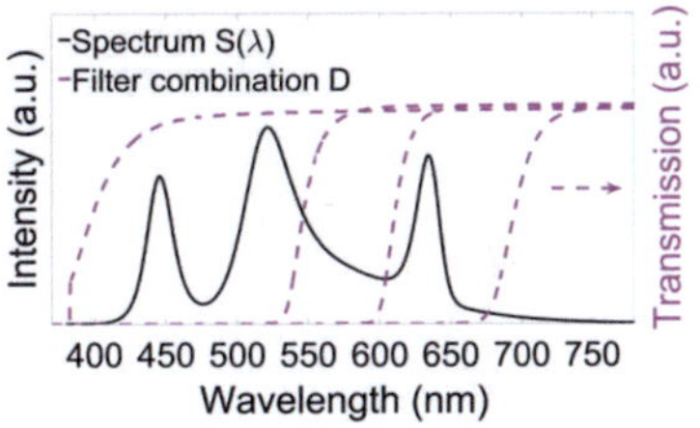

(d) Filter combination D

Figure 6.1: Reconstruction of a RGBW LED: According to the reconstruction Eq. 4.16, a spectra consisting of n_{max} known basis spectra can generally be reconstructed with n_{max} arbitrary filter measurements and would therefore work equally well for all displayed filter combinations.

The uncertainty η_n from Eq. 6.1 has two different physical origins as shown in Eq. 6.2.

$$M_n + \eta_n = (M_{\mathrm{t},n} + \eta_{\tau,n}) \times (1 + \eta_{\mathrm{N}}) \tag{6.2}$$

The first factor consists of uncertainties $\eta_{\tau,n}$ regarding the effective spectral sensitivities $s_{\tau,\mathrm{eff},n}(\lambda)$ and their interactions with the basis spectra. The uncertainty $\eta_{\tau,n}$ depends on the individual filter used in the measurement and is added to the true value $M_{\mathrm{t},n}$. The second term η_{N} describes uncertainties which apply for all filter measurements independent of the specific filter function $s_{\tau,\mathrm{eff},n}(\lambda)$ at each step of the goniophotometric measurements. Examples are noise from the sensor chip or the quantification errors caused by the ray file resolutions. Furthermore, the uncertainties propagate while solving the reconstruction Eq. 4.17. This error propagation can be estimated with the condition number of the reconstruction matrix $M_{S\mathcal{T}}$ and according to Eq. 6.3 [88],

$$\frac{\|\Delta A\|_2}{\|A\|_2} = \underbrace{\left(\left\|M_{S\mathcal{T}}^{-1}\right\|_2 \cdot \|M_{S\mathcal{T}}\|_2\right)}_{\mathrm{cond}_2(M_{S\mathcal{T}})} \frac{\|\eta\|_2}{\|M\|_2} \tag{6.3}$$

in which the variables ΔA, A, η and M are each a vector corresponding to a set of $n_{\max}$ measurements. Note that Eq. 6.3 is suitable as a rough estimation, especially if $\mathrm{cond}_2(M_{S\mathcal{T}})$ is very large. But it is not capable of distinguishing between the different amplitude values ΔA_n and is therefore not suited to judge the performance of a reconstruction in detail. A detailed estimation of the reconstruction performance of a filter combination needs to consider the individual filter based η as well as its error propagation. Thus the perfomance will differ for each filter combination and selecting a robust filter combination becomes an essential part of applying PMBS.

In the following sections, empirical models for optical glass and common interference filters as they are provided by the manufacturers are extended to describe the physically caused measurement uncertainty $\eta_{\tau,n}$. The models allow the analysis of the influence of different effective spectral sensitivities $s_{\mathcal{T},\mathrm{eff},n}(\lambda)$ on the reconstruction result of PMBS. This offers the possibility of choosing the optimal filter combination from a given filter set.

6.1.1 Optical glass filters

An optical glass filter is characterized by its internal transmission spectrum $\tau_{\mathrm{in}}(\lambda)$ given at a reference thickness d_0. The ideal transmission function $\tau(\lambda)$ is the product of the resulting internal transmission $\tau_{\mathrm{in}}(\lambda)$ for a thickness d and the Fresnel losses $\tau_{\mathrm{Fr}}(\gamma, n_{\mathrm{r}})$ at the air/glass and glass/air interfaces, which depend on the angle of incidence γ and the refractive index n_r. The ideal relation is given in Eq. 6.4 [89].

$$\tau(\lambda) = \tau_\mathrm{Fr}(\gamma, n_\mathrm{r}) \times \tau_\mathrm{in}(\lambda)^{\frac{d}{d_0}} \tag{6.4}$$

The uncertainties of the room temperature $u(T)$ and temperature shifts $u(T_\mathrm{abs})$ due to absorption as well as those of spatial roughness $u(d)$, angle of incidence $u(\gamma)$, the uncertainty of the transmission curve measurement $u(\tau_\mathrm{in})$ and its wavelength precision $u(\lambda)$ [84] are added to the model as absolute values. They influence Fresnel losses as well as the internal transmissions. All uncertainties are assumed to be correlated and are therefore the same for all wavelengths. Note that all uncorrelated uncertainties of $\tau(\lambda)$ are averaged out due to the spectral integration on the ILMD sensor. The model of the transmission function therefore depends on the angle of incidence γ, the refractive index n_r, the thickness of the glass d and the temperature coefficient $\mathrm{d}\lambda/\mathrm{d}T$. Hence the model is defined as

$$\begin{aligned}\tau(\lambda) &= \tau_\mathrm{Fr}(\gamma + u(\gamma), n_\mathrm{r}) \\ &\times \left[u(\tau_\mathrm{in}) + \tau_\mathrm{in}\left(\lambda + u(\lambda) + \frac{\mathrm{d}\lambda}{\mathrm{d}T}\left[u(T) + u(T_\mathrm{abs})\right]\right)\right]^{\frac{d+u(d)}{d_0 \cos\delta}}\end{aligned} \tag{6.5}$$

using

$$\delta = \arcsin\left\{\frac{\sin\left[\gamma + u(\gamma)\right]}{n_\mathrm{r}}\right\}.$$

Multiple reflections in the Fresnel term $\tau_\mathrm{Fr}(\gamma + u(\gamma), n_\mathrm{r})$ are not modeled since the absolute difference between multiple reflections and a two boundary approach is small. By neglecting the effect in the model this small absolute difference is not assigned to $\tau_\mathrm{Fr}(\gamma + u(\gamma), n_\mathrm{r})$[2]. For

[2] If for instance the refractive index of the optical glass is 1.56, the difference of the transmission at normal incidence would be smaller than 0.25%. Therefore, by not considering the effect as a functional relation, merely the change of those 0.25% difference as a function of the uncertainty terms is neglected.

the same reason, dispersion effects and thermal expansion [86] are neglected as well. Also note that the linear temperature dependence in Eq. 6.5 is limited to long pass glass edge absorption filters [89].

6.1.2 INTERFERENCE FILTERS

The transmission spectrum $\tau(\lambda)$ of interference filters relies on constructive and destructive interference at a stack of specially designed interfaces. They can be designed with different FWHM (Full Width at Half Maximum) at different center wavelengths. The sources of uncertainty are the same as for the glass filters despite heating due to absorption and spatial roughness. The shift of the transmission spectrum can be estimated for small angles ($\gamma < 15°$), where n_{eff} is the effective refraction index of the spacer layer and λ the wavelength of the shifted spectral feature [89]–[92]. Decreasing transmission, the broadening of the transmission spectrum and a separation of s- and p-polarization can be neglected for the angle of incidence used in this thesis [91]. The model of the interference filters for small angles $\gamma < 15°$ can therefore be written as

$$\tau(\lambda) = u(\tau_{\text{in}}) + \tau_{\text{in}} \left(\lambda + u(\lambda) + \frac{\mathrm{d}\lambda}{\mathrm{d}T} u(T) + u(\lambda_\gamma) \right) \tag{6.6}$$

using

$$u(\lambda_\gamma) = \lambda \sqrt{1 - \frac{1}{n_{\text{eff}}} \sin\left[\gamma + u(\gamma)\right]^2}.$$

6.1.3 INPUT UNCERTAINTIES

To take into account the influence of the uncertainty terms in Eq. 6.5 and 6.6, the uncertainties need to be modeled as well. In accordance with [93], a probability density function (PDF) is chosen or determined for each term. Table 6.1 summarizes the accounted uncertainty terms and their model parameters.

The noise from the sensor chip η_N increases linearly with the signal level of an individual pixel, if the dark signal noise, which is just in the order of a few least significant bits, is neglected [94]. The term η_N is assumed to be a Gaussian. The absolute values are in the order of the repeatability of a commercial luminance/color camera [58]. The quantification error of the ray file is neglected as it depends on the measurement resolutions, the number of rays and the model resolutions used to apply PMBS.

The uncertainty of the internal transmission wavelength $u(\lambda)$ is also modeled as a Gaussian distribution. The same is true for the uncertainty of the spatial roughness $u(d)$, which influences the thickness of the filter across its area. The room temperature uncertainty $u(T)$ is assumed to be sinusoidal, because the room temperature is usually controlled to lie in a given interval. The measurement movements of the goniometer justify a similar assumption for the temperature uncertainty caused by absorption $u(T_{abs})$. The uncertainty of the optical transmission amplitudes and especially the blocking range $u(\tau_{in})$ are assumed to lie in the interval provided by the manufacturers and are therefore modeled as a rectangular distribution.

The most complex term is the angular distribution because it depends on the geometric setup as well as the source and its radiance distribution. In this thesis, it is modeled as an uncertainty $u(\gamma)$ as well. It bases solely on geometrical assumptions regarding the diameter of a circular light source with a constant radiant intensity, a circular filter diameter and the measurement distance. The filter is assumed to be attached in front of the lens of the ILMD. Figure 6.2 (a) shows the obtained distribution for a point source and a small (20 mm) and a large (70 mm) light source with a filter diameter of 40 mm and a measurement distance of 300 mm. The geometrical assumption is shown in Fig. 6.2 (b).

Table 6.1: Model parameters of the considered uncertainties

Term	PDF	Amplitude
$u(\tau_{\text{in}})$	Rectangular	10^{-5}
$u(\lambda)$	Gaussian	0.5 nm
$u(d)$	Gaussian	3 μm
$u(T)$	Sinusoidal	3 K
$u(T_{\text{abs}})$	Sinusoidal	3 K
η_{N}	Gaussian	0%
		0.4%
$u(\gamma)$	see Fig. 6.2	small
		large

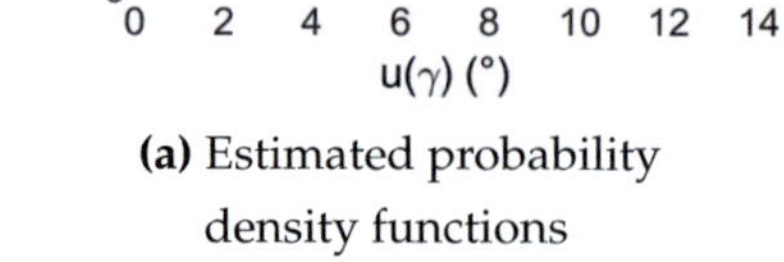

(a) Estimated probability density functions

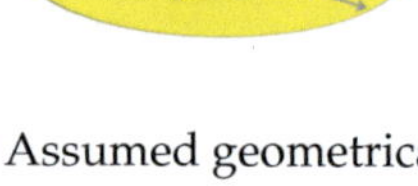

(b) Assumed geometrical setup

Figure 6.2: Estimation of angle of incidence $u(\gamma)$ on filter.

6.2 FILTER SELECTION

As described above, the reconstruction result is influenced by the optical filters of the ILMD-based measurements. Due to the fact that these measurements are a time-consuming aspect of PMBS, it is not feasible to try out all different filters in real measurements. Therefore, the ideal set of $n_{\max}$ optimal filters from all possible filter combinations has to be determined prior to the actual measurements. Given a set of m different filter functions $\tau(\lambda)$, one can determine the ideal subset of $n_{\max}$ filters out of N_{comb} possible filter combinations for each specific $R(X, Y, Z, \phi, \vartheta, \lambda)$ consisting of $n_{\max}$ basis spectra $S_n(\lambda)$. N_{comb} can be calculated with the binomial coefficient:

$$N_{\text{comb}} = \begin{pmatrix} m \\ n_{\max} \end{pmatrix} = \frac{m!}{n_{\max}!(m-n_{\max})!} \tag{6.7}$$

The following subsections introduce two methods to determine the optimal filters by considering the modeled uncertainties. Figure 6.3 shows the workflow of PMBS under consideration of both methods. The first method uses Monte Carlo simulations to determine the ideal filter combination from a given filter set. The second - optional and less accurate, but faster - method which is based on factorial design, may be used prior to the actual Monte Carlo simulation to reduce the number of all possible combinations to save computation time.

6.2.1 MONTE CARLO BASED FILTER SELECTION

To determine the ideal filter combination a Monte Carlo simulation can be performed. In this simulation, random amplitudes A_n of the basis spectra are generated to create a defined number of known random spectra $S(\lambda)$. Furthermore, random uncertainty values are created by using inverse transform sampling of the predefined PDFs in table 6.1. The uncertainties are sampled individually for each filter.

Subsequently, the measurements are simulated by solving Eq. 4.15 using the known random spectra amplitudes A_n and the transmission profiles, which are influenced by the uncertainty terms according to Eq. 6.5 and 6.6. This results in a simulated measurement value $M_{\eta,n} = M_n + \eta_n$ for each filter for each random spectrum. Then the sensor noise is added according to Eq. 6.2. Finally the reconstruction equation Eq. 4.17 is performed to obtain $A_{\eta,n} = A_n + \Delta A_n$. These $A_{\eta,n}$ define the reconstructed spectrum $S_\eta(\lambda)$, which can be compared to the known random spectrum $S(\lambda)$ with the residual sum of squares RSS according to

$$RSS = \sum_i \sum_\lambda k_i(\lambda) \times \left(S_\eta(\lambda) - S(\lambda)\right)^2 . \tag{6.8}$$

One or several weighting functions $k_i(\lambda)$, for example the color matching functions, can be added to rate the RSS according to the desired application. The mean value of all random spectra RSS_{mean} is used as a figure of merit for the specific filter combination. The mean value is chosen because it includes outliers stronger than the median, which would overestimate the filter performance. Note that choosing the maximum would base the selection on too noisy data. The filter combination minimizing RSS_{mean} would be the preferred one for the goniophotometric measurements and the spectral reconstruction. Since the absolute values of RSS_{mean} do not allow an intuitive judgment of the colorimetric precision, the known random spectra and the reconstructed spectra can also be compared in terms of absolute colorimetric distance $\triangle u'v'_{\text{mean}}$. As described in chapter 4, $\triangle u'v'_{\text{mean}}$ alone is not suitable to evaluate the performance of the reconstruction, if there are more than three basis spectra, due to metamerism.

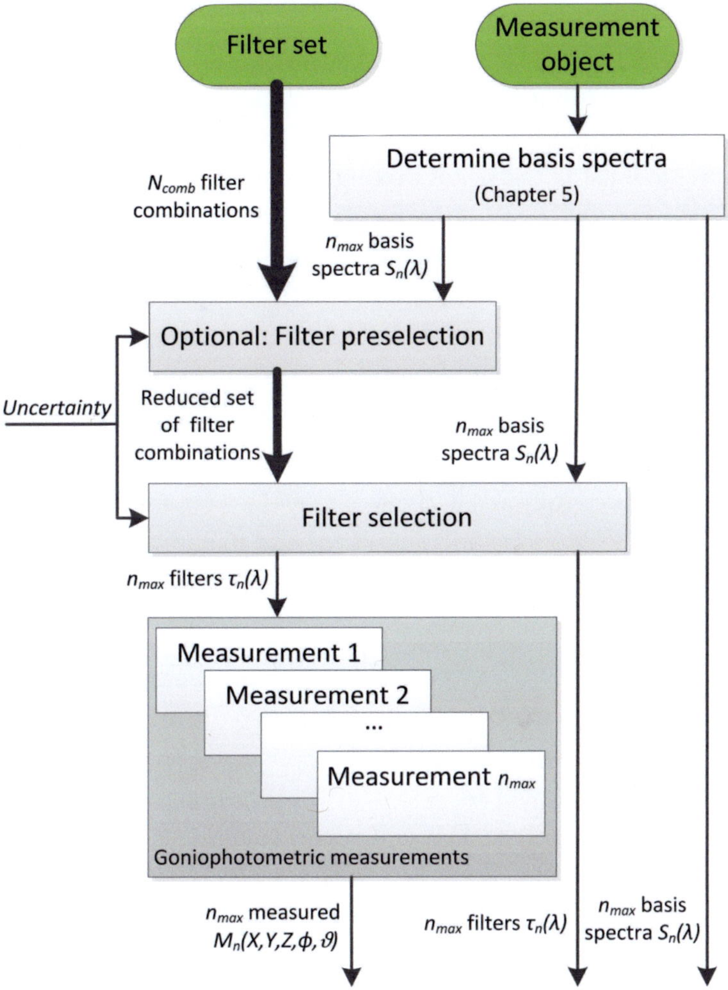

Figure 6.3: Integration of filter selection in workflow of PMBS.

6.2.2 FACTORIAL DESIGN BASED FILTER PRESELECTION

An individual Monte Carlo simulation for each possible filter according to Eqs. 4.14, 6.2, 6.5, and 6.6 and each possible combination according to Eq. 4.17 can result in a high computation time for the filter selection process because the number of combinations can become very large. Hence, a preselection method to reduce the number of filters and filter combinations to be addressed by the Monte Carlo simulations is developed.

Firstly, a two level full factorial design [95] is used to estimate the physical filter errors according to Eqs. 4.14, 6.5 and 6.6 for a specific test spectrum. A two level full factorial design describes the permutation of chosen minimal and maximal values for all m input factors resulting in 2^m iterations. As minimal and maximal factors of each uncertainty the terms

$$\begin{aligned} u_{\text{min}} &= u_{\text{mean}} - u_{\sigma} \\ u_{\text{max}} &= u_{\text{mean}} + u_{\sigma} \end{aligned} \tag{6.9}$$

are defined. They are the difference between the uncertainty factors from their mean value u_{mean} and the standard deviation u_{σ} derived from their PDFs. There are $m = 6$ factors resulting in 64 iterations for each optical glass filter and $m = 4$ factors resulting in 16 iterations for each interference filter according to Eqs. 6.5 and 6.6. If more detailed models which account for more factors than Eq. 6.5 and 6.6 are used the number of iterations can be reduced by using a fractional factorial design [96]. Subsequently, the standard deviation is calculated based on all iterations to represent $\eta_{\tau,n}$ for each individual filter.

After estimating the uncertainty of each filter, the error propagation resulting from the solution of Eq. 4.17 has to be considered. Therefore, the second part of the preselection process estimates the mathematical error of each filter combination with a 3-level full factorial design

[95]. The sensor signal noise η_{N} is added to the results of the first simulation according to Eq. 6.2. This results in a minimal, mean and maximal value for each filter:

$$\begin{aligned} M_{\mathrm{min},n} &= (M_{\mathrm{t},n} - \eta_{\tau,n}) \times (1 - \eta_{\mathrm{N}}) \\ M_{\mathrm{mean},n} &= M_{\mathrm{t},n} \\ M_{\mathrm{max},n} &= (M_{\mathrm{t},n} + \eta_{\tau,n}) \times (1 + \eta_{\mathrm{N}}) \end{aligned} \tag{6.10}$$

Note that $M_{\mathrm{mean},n}$ represents the true value $M_{\mathrm{t},n}$ without any uncertainty. The test spectrum $S(\lambda)$ is reconstructed for all $3^{n_{\mathrm{max}}}$ possible permutations for each filter combination by solving the reconstruction equation Eq. 4.17 to define $S_{\eta}(\lambda)$. Again, the number of iterations can be reduced by either using a two level factorial design, a fractional factorial design or both. As in the Monte Carlo simulation, the RSS_{mean} between the original spectrum $S(\lambda)$ and the $3^{n_{\mathrm{max}}}$ reconstructed spectra $S_{\eta}(\lambda)$ is used as figure of merit. After selecting a certain percentage of the N_{comb} possible combinations which performed best in the preselection process, the full Monte Carlo simulation is performed to select the optimal combination from the reduced sub set.

6.3 VALIDATION

The Monte Carlo simulation has to determine the best or at least one of the best filter combinations such that only the required ILMD-based goniophotometric measurements are performed. The filter preselection has to determine the filter sets such that the most promising filter combinations are not excluded by the preselection. This section aims to validate these requirements.

6.3.1 VALIDATION OF FILTER SELECTION

To validate the filter selection, simulations are compared to spectral reconstructions, which are based on measurements with different filters. The filter selection is more important and more complex in the case of a higher number of overlapping basis spectra. Therefore, a RGBW LED system was used to verify the filter selection process. To minimize the influences of the basis spectra modeling, the basis spectra are based on individual spectral measurements of the red, green, blue and white LEDs. The basis spectra are displayed in Fig. 6.4 (a).

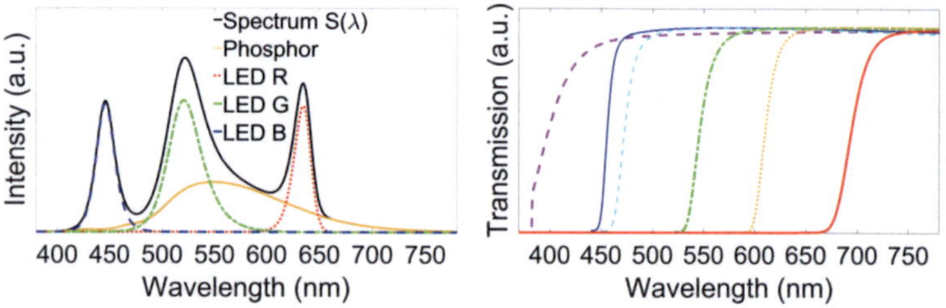

(a) Modeled basis spectra

(b) Glass absorption filters

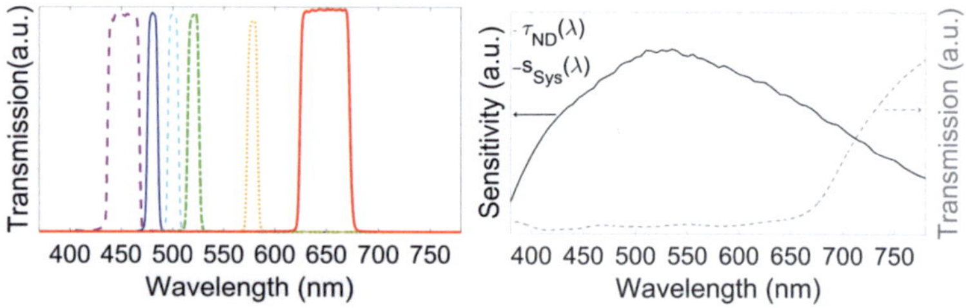

(c) Interference filters

(d) CCD sensor and ND filter

Figure 6.4: Validation of filter selection: Data of the simulation.

The filter selection process was performed to estimate the arrangement of the 15 possible filter combinations from ILMD-based measurements with six edge absorption filters as well as from ILMD-based measurements with six interference filters as displayed in Fig. 6.4 (b) and (c). The Monte Carlo simulation also included the spectral sensitivity of the remaining system as well as the transmission function of a neutral density filter which had to be added to all ILMD-based measurements in order to reduce the signal on the CCD. Both are displayed in Fig. 6.4 (d). The uncertainties are based on table 6.1 with $\eta_\mathrm{N} = 0.4\%$ and a large angular distribution. The neutral density filter transmission uncertainty was taken into account and modeled similarly to $u(\tau_\mathrm{in})$ and $u(\lambda)$ but only once for all filters as the same neutral density filter was used in each measurement.

In addition to the twelve ILMD-based goniophotometric measurements, an angularly resolved spectral measurement was performed. The angularly resolved spectral measurement is used to evaluate the performance of each filter combination. The evaluation used is similar to the punctual spectral measurements described in subsection 5.3.2. Each spectrum is reconstructed at the known angular positions and compared to the measured spectrum in terms of the chromaticity distance $\triangle u'v'$ and the RSS. The mean value of all positions RSS_mean is used to rate each filter combination. The mean chromaticity distance $\triangle u'v'_\mathrm{mean}$ is used to assess the absolute influence of the filter selection with respect to the requirements of the spectral ray files.

Figure 6.5 shows the distribution of the RSS obtained for the four filter combinations A-D from the beginning of this chapter. The logarithmic scale is normalized to the worst RSS of all reconstructions. As suspected, the simulation hints a different reconstruction performance for the four filter combinations A-D.

Figure 6.6 shows the obtained $\triangle u'v'_\mathrm{mean}$ as a function of the selection number of the filter selection. The reconstruction performance of the glass edge absorption filters in Figure 6.6 (a) can be divided into roughly three groups. The first group results in a $\triangle u'v'_\mathrm{mean} \approx 0.01$, the second group in a $\triangle u'v'_\mathrm{mean} \approx 0.025$ and the last group in a

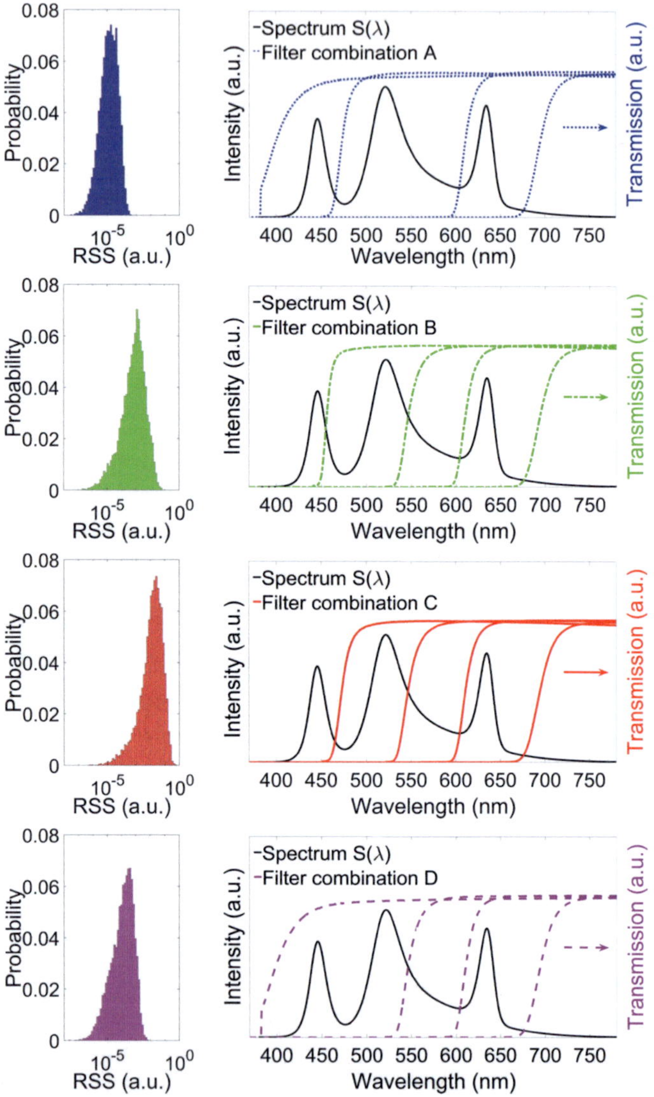

Figure 6.5: Validation of filter selection: Comparison of the simulated reconstruction results of four different filter combinations

$\triangle u'v'_{\text{mean}} > 0.1$. The last group consists of filter combinations which cannot reconstruct the spectra due to missing information such as filter combination B and C. Impossible chromaticity distance values such as $\triangle u'v'_{\text{mean}} > 0.6$ result from physically impossible chromaticity coordinates, which are based on partly negative spectral distributions. The impossible area is marked in all figures. The filter selection succeeds in separating these three groups. The order within the groups is not monotone but particularly in the first group the differences are very small. Nevertheless, the filter selection identifies the optimal filter combination, which is filter combination A, from all 15 possibilities.

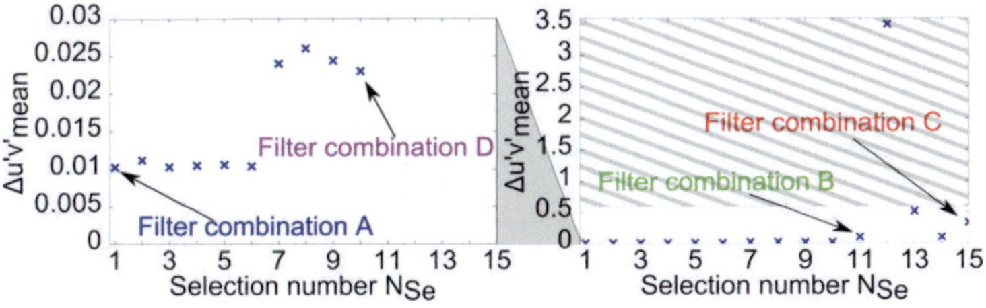

(a) Edge absorption filters

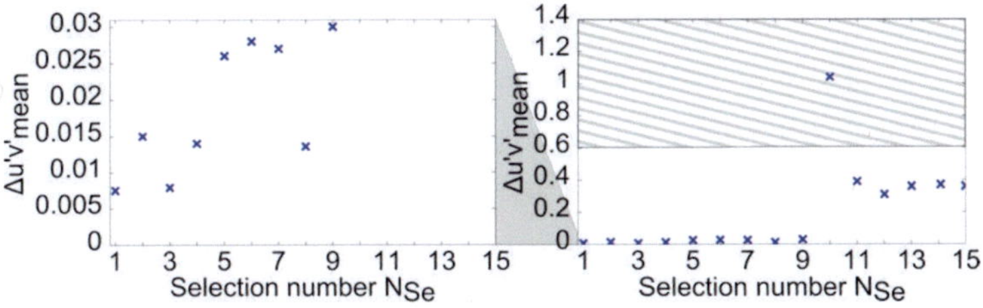

(b) Interference filters

Figure 6.6: Validation of filter selection: Experimental reconstruction results as a function of filter selection number (marked area contains unphysical $\triangle u'v'_{\text{mean}}$ due to negative spectral parts).

Similarly to the glass edge absorption filters, the performance of the interference filters in Fig. 6.6 (b) can be divided into groups. The first group results in a reconstruction performance of $\triangle u'v'_{\mathrm{mean}} < 0.01$. The second results in $\triangle u'v'_{\mathrm{mean}} \approx 0.015$, the third in $\triangle u'v'_{\mathrm{mean}} \approx 0.025$ and the last group in $\triangle u'v'_{\mathrm{mean}} > 0.1$. While the selection number arrangement still represents the reconstruction well, there seem to be outliers at the second/third and eighth position. On closer inspection, the reconstructions with the better performance at the first, third and eighth position all exclude the interference filter with the center wavelength around 480 nm. All other filter combinations in Fig. 6.6 (b) include this specific filter. There are two possible explanations for this behavior. The first one is neglected uncertainties, such as those of the CCD, whose spectral sensitivity function might affect the interference filter at 480 nm more than the alternative filters at 500 nm and 520 nm. But it is more likely that the disparity is caused by deviations from the basis spectra model of the blue LED and the phosphor to the real basis spectra. This difference is stronger in the spectral region in which both spectra overlap. Since the transmission region of the 480 nm filter is located directly in this spectral region, the assumed reconstruction matrix might differ from the true matrix. The filter selection cannot address this circumstance, because it assumes that the basis spectra are correct. This issue will be discussed in more detail in section 6.4. Nevertheless, the filter selection of the interference filters also identifies the best filter combinations from all possible combinations.

6.3.2 VALIDATION OF FILTER PRESELECTION

Similarly to the Monte Carlo simulation, the result of the preselection is an arrangement of all possible filter combinations according to their anticipated reconstruction success. To validate the performance of the preselection, the resulting arrangement is compared to the results of the Monte Carlo simulation.

Therefore, simulations with four different test spectra as shown in Fig. 6.7 have been performed. The test spectra include the combination of Blue/Yellow (BY), Red/Blue/Yellow (RBY), Blue/Blue/Yellow (BBY) and Red/Green/Blue/Yellow (RBGY). The evenly distributed filter set used in this analysis consists of 63 interference filters (three with 1.5 nm, 47 with 10 nm, seven with 40 nm and six with 70 nm FWHM) from [90], [97] and 23 glass filters in the visible spectral range. The glass filters consist of the 18 glass edge absorption filters GG380-RG715 and five band pass filters from [89]. Typical internal transmission profiles, temperature coefficients and refractive indices were provided by the manufacturers. Note that in this example 86 filter transmission profiles led to over 2×10^6 combinations for test spectrum RGBY ($n_{\max} = 4$).

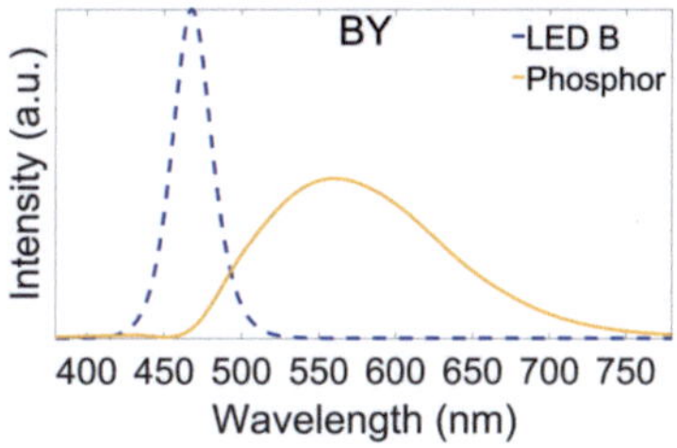

(a) Blue/Yellow

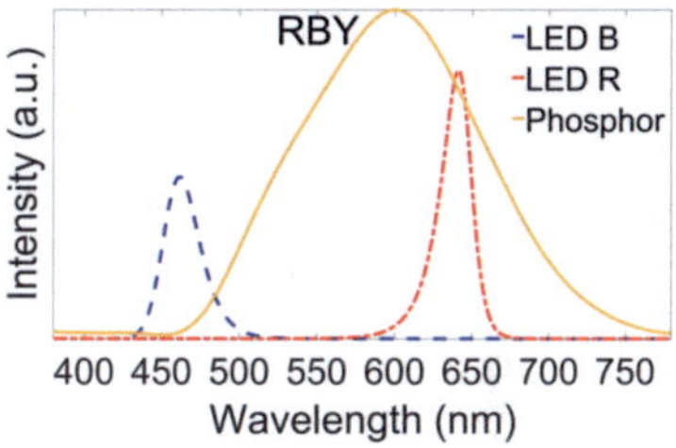

(b) Red/Blue/Yellow

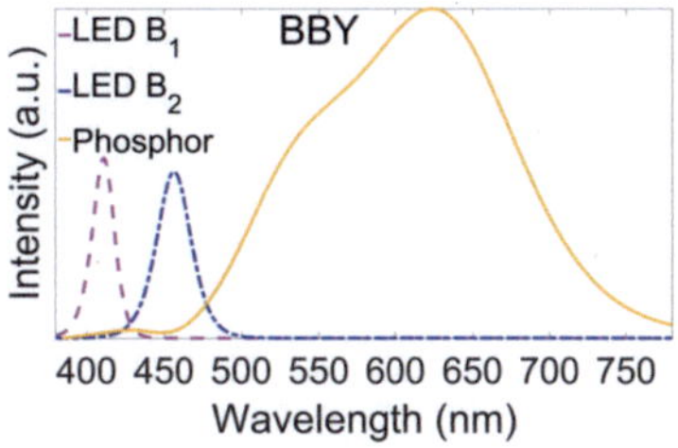

(c) Blue/Blue/Yellow

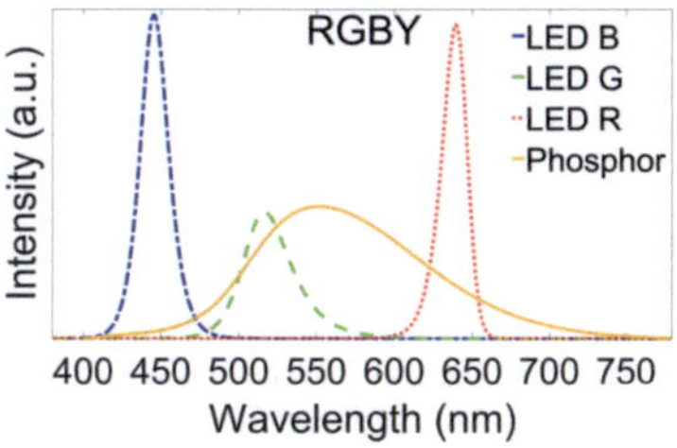

(d) Red/Green/Blue/Yellow

Figure 6.7: Validation of preselection: Visualization of the four test spectra.

The uncertainty terms were modeled according to table 6.1, such that realistic values are used for this verification. The simulation included a constant system function, which described the sensitivity of a CCD sensor without an uncertainty and without any neutral density filter. As the angular distribution varies strongly depending on the geometric extension, the angular uncertainty distribution was modeled twice as shown in Fig. 6.2 to present a large and small measurement object. The same is true for the sensor noise and the quantification error of the ray files η_{N}. Therefore, the total number of filter selection scenarios adds up to 16.

For each of the 16 selection scenarios, a Monte Carlo simulation and a preselection have been performed. For each Monte Carlo simulation, 10^5 random amplitude sets have been created. The amplitude sets have been modeled as rectangular distribution within a high dynamic range of 140 dB [58]. The amount of modeled filter combinations in the Monte Carlo simulation consists of the first 50% possible filter combinations according to the preselection for test spectra BY, RBY and BBY and 10% for test spectrum RGBY since the large number of possible combinations results in a high computation time.

Figure 6.8 shows four exemplary results from the 16 simulations. The mean residual sum of squares RSS_{mean} obtained from the Monte Carlo simulation of each filter combination is plotted on a logarithmic scale against the preselection number N_{PS} to check the validity of the preselection process. A perfect preselection would be a monotonically increasing graph. However, in practical applications, it is most important that the ideal filter combination, or at least a combination which is nearly as good as the ideal one, is within the first part of the preselection. There must not be strong outliers towards small values at late preselection numbers since this would prohibit a more detailed analysis in the real Monte Carlo simulation. However, an overestimation of bad filter combinations is uncritical since it will be detected during the succeeding Monte Carlo simulation. The highlighted areas in Fig.6.8 correspond to the first two percent of all possible filter combinations. All most promising combinations including the global minimum of RSS_{mean} are within this area.

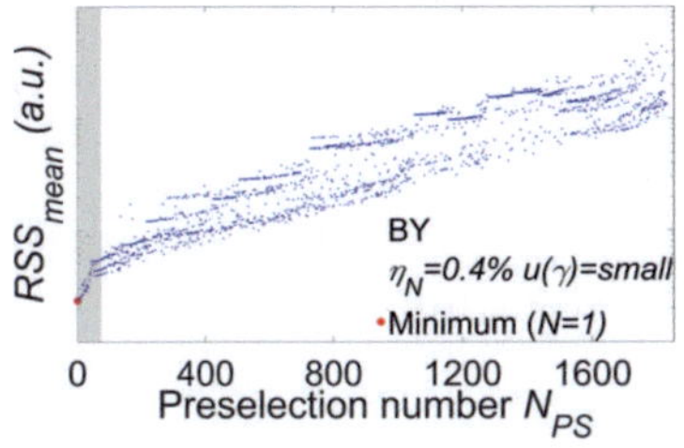

(a) Spectrum BY

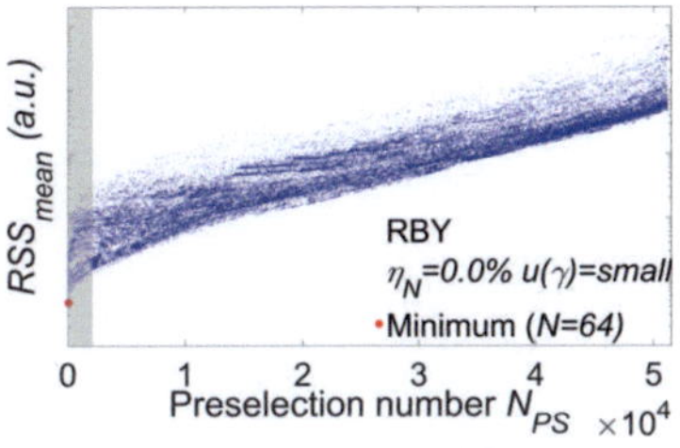

(b) Spectrum RBY

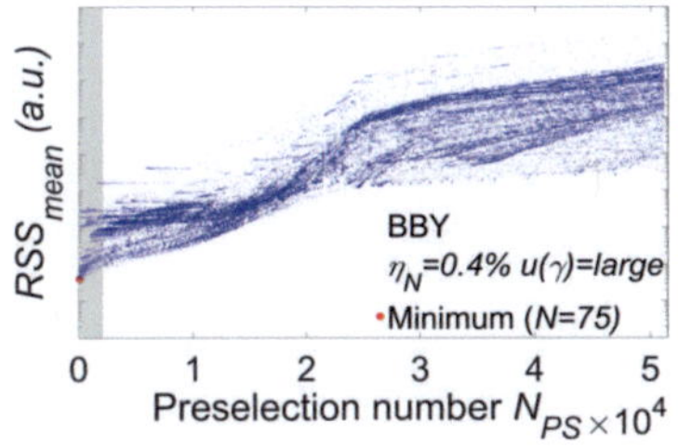

(c) Spectrum BBY

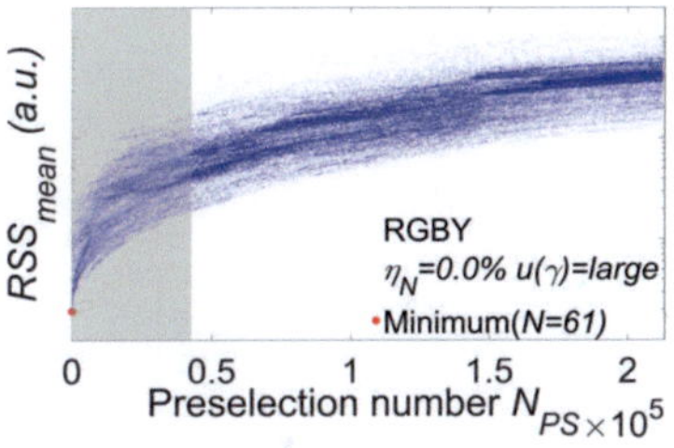

(d) Spectrum RGBY

Figure 6.8: Validation of preselection: The reconstruction performance RSS_{mean} is plotted on a logarithmic scale against the preselection number N_{PS} of different filter combinations for all four test spectra.

The preselection number N_{PS} of the ideal filter combination and its relative position $\frac{N_{\text{PS}}}{N_{\text{comb}}}$ from all possible combinations in percent for all 16 selection scenarios are summed up in table 6.2. All relative positions are within the first few percent of the possible combinations. Note that the results in percent are more critical for fewer possible combinations. However, one can for example save a factor 50 of computation time by simply simulating the best 2% from the preselection.

Table 6.2: Validation of preselection: The table shows the preselection number N_{PS} of the optimal combination and its relative position $\frac{N_{PS}}{N_{comb}}$ for all 16 selection scenarios.

Spectrum	$u(\gamma)$	η_N	Results preselection	
			N_{PS}	Relative position $\frac{N_{PS}}{N_{comb}}$
BY	small	0.0%	1	0.03%
		0.4%	1	0.03%
	large	0.0%	1	0.03%
		0.4%	1	0.03%
RBY	small	0.0%	64	0.06%
		0.4%	234	0.23%
	large	0.0%	79	0.08%
		0.4%	87	0.09%
BBY	small	0.0%	89	0.09%
		0.4%	34	0.03%
	large	0.0%	56	0.05%
		0.4%	75	0.07%
RGBY	small	0.0%	59	<0.01%
		0.4%	2	<0.01%
	large	0.0%	61	<0.01%
		0.4%	2	<0.01%

6.4 FILTER TECHNOLOGY COMPARISON

The validation shows that the filter selection as well as the preselection fulfill their requirements. But it also shows that there is no significant difference regarding the reconstruction performance of the two different filter technologies. In general, the impact of uncertainties affecting the wavelength precision of the transmission function $\tau(\lambda)$ strongly depends on the spectral position of steep slopes. If the filter and a basis spectrum both have a steep slope in the same spectral region, the impact on the measurement result, and therefore the relative importance of those uncertainties, increases. An important difference between the technologies is that glass absorption filters are more strongly affected by temperature variations and that the interference filters are more strongly affected by the angular distribution, which depends on the device under test. Furthermore, the glass edge absorption filters tend to set up reconstruction matrices which propagate errors more strongly since they provide a higher condition number. If the basis spectra are certain, there is no important difference between both technologies.

However, the validation of the Monte Carlo simulation shows another important aspect of the filter selection, which is the ignored uncertainty of the modeled basis spectra. This uncertainty may change across the wavelengths, as in the case of the phosphor and the blue LED. If a filter only transmits spectral regions which are uncertain with respect to the basis spectra, the spectral reconstruction deteriorates. As described in section 5.3, a basis spectra optimization is required to separate complex spectra such as GBY (GW) or RGBY. However, as this optimization relies on ILMD-based measurements, the filter selection has to take place prior to the basis spectra optimization if no additional measurements are to take place. Consequently, the filter selection should remain stable in the case of inaccurately estimated basis spectra, as for instance those which are based solely on the initial modeling of section 5.2.

In this context, filters which provide a larger transmitting region are expected to perform better. These would be glass edge absorption filters. To briefly validate this expectation, the basis spectra of the RGBW LED were estimated only with the initial modeling of 5.2 and the selection number was obtained with these estimated basis spectra. The numbers in table 6.3 and 6.4 show the original selection numbers, which are based on the individually modeled basis spectra. The row numbers, which are equal to the selection numbers of the first column, stand for the selection numbers of the inaccurately estimated basis spectra for model 1-4. The measured reconstruction performance is coded in the cell color according to Fig. 6.9. If the cell color is green, the performance was within the best group, if it is yellow within the second best group and if it is red it failed to reconstruct the spectrum. If the cell color within a row does not change, the filter selection can be assumed to be stable. Table 6.3 shows that the filter selection based on the glass edge absorption filters is quite stable. Although there is a change of sequence within the second group and in the case of model 3 also a change between groups, the filter selection stays stable.

Table 6.4 and Fig. 6.10 show that the filter selection which is based on the interference filters has more sequence changes between the groups and in one instance even fails to select the optimal filter combination. For a fair comparison to the glass edge absorption filters, the blue cell colors have to be interpreted as either green or yellow. Nevertheless, it can be concluded that in the case of uncertain basis spectra the transmission region of the filters should remain large, which is a more typical attribute of glass edge absorption filters.

Table 6.3: Filter selection number with inaccurately estimated basis spectra using glass edge absorption filters

True basis spectra	Model 1	Model 2	Model 3	Model 4
1	1	1	1	1
2	2	2	2	2
3	3	3	3	3
4	4	4	4	4
5	5	5	5	5
6	6	6	6	6
7	10	10	10	10
8	9	7	14	7
9	7	8	13	8
10	8	9	7	9
11	11	11	8	14
12	13	13	9	13
13	12	12	11	11
14	14	14	12	12
15	15	15	15	15

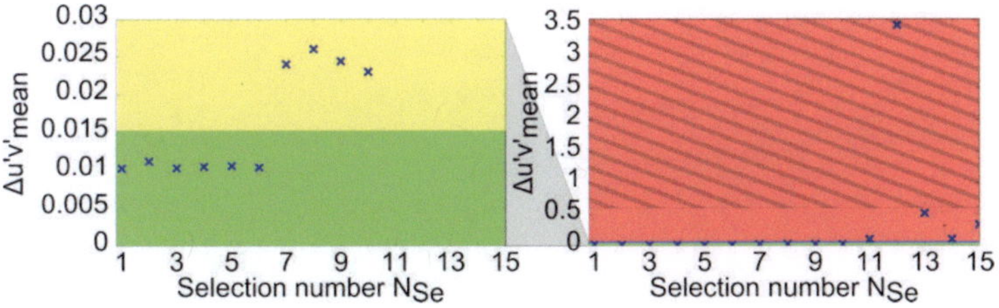

Figure 6.9: Reconstruction success of true basis spectra for edge absorption filters and allocation to different groups (marked area contains unphysical $\triangle u'v'_{\mathrm{mean}}$ due to negative spectral parts).

Table 6.4: Filter selection number with inaccurately estimated basis spectra using interference filters

True basis spectra	Model 1	Model 2	Model 3	Model 4
1	1	1	2	1
2	3	3	1	2
3	2	2	5	5
4	5	4	8	4
5	4	5	4	3
6	8	7	3	8
7	7	8	7	7
8	6	6	6	6
9	12	9	11	9
10	11	10	12	11
11	10	15	9	10
12	9	11	15	12
13	15	14	14	15
14	13	12	10	14
15	14	13	13	13

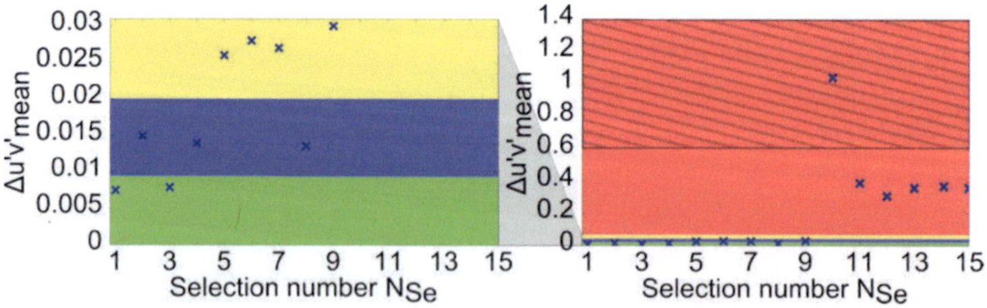

Figure 6.10: Reconstruction success of true basis spectra for interference filters and allocation to different groups (marked area contains unphysical $\triangle u'v'_{\text{mean}}$ due to negative spectral parts).

CHAPTER 7

PMBS: CREATION OF HYPERSPECTRAL LED MODELS

As discussed in chapter 4, the generation of spectral ray files using PMBS requires solving a system of linear equations by applying the reconstruction equation. However, problems occur due to the resolution and precision requirements of spectral ray files and the discrete nature of the input data. This chapter introduces histogram based light source models to apply PMBS. It focuses on the relationship of artifacts and resolutions and introduces a physically motivated calculation concept using neighborhood relations to overcome the highly discrete nature of ray files, which finally leads to spectral ray files [1].

7.1 CALCULATION CONCEPT

Theoretically, the general problem of obtaining spectral ray files is solved by an adequate basis spectra modeling to obtain $S_n(\lambda)$ as described in chapter 5 followed by an optimal filter selection to determine the optimal $s_{\tau,\text{eff},n}(\lambda)$ as described in chapter 6. Then it is only necessary to apply the reconstruction Eq. 4.17 on the 5-dimensional measured ray files. However, due to the discrete nature of ray files, the rays of different ray files will not be defined at the same positions (X, Y, Z) with the same directions (ϕ, ϑ). To this end the calculation concept as shown in Fig. 7.1 has been developed.

[1] *Parts of this chapter have already been pubslished in [98].*

The calculation process starts with the transformation of the measured ray files into continuous models of their plenoptic function $M_n(X, Y, Z, \phi, \vartheta)$. Subsequently, Eq. 4.17 is applied to the measurement models to achieve the spectral models $A_n(X, Y, Z, \phi, \vartheta)$. Finally, the spectral models can be used to create spectral ray files.

Section 7.2 describes the transformation process from a ray file into a continuous histogram based measurement model and vice versa. Furthermore, it describes the model precision parameter NCC (normalized cross-correlation) necessary to validate the model resolution. The third section 7.3 focuses on the spectral models. It is shown that the application of the reconstruction equation Eq. 4.17 on high precision models regarding the NCC can lead to insufficient spectral reconstruction results. Therefore, a second precision parameter, which is the physically motivated amount of negative amplitudes, is derived to validate the spectral models. Furthermore, a post-processing step, which utilizes neighborhood relations in the spectral models and minimizes these reconstruction errors, is introduced. This improvement enables the derivation of the final workflow in section 7.4, which combines high model resolutions and high spectral accuracies by taking into consideration both defined precision parameters and their relations. The last section includes the validation of this chapter based on a spatially and spectrally well defined test device.

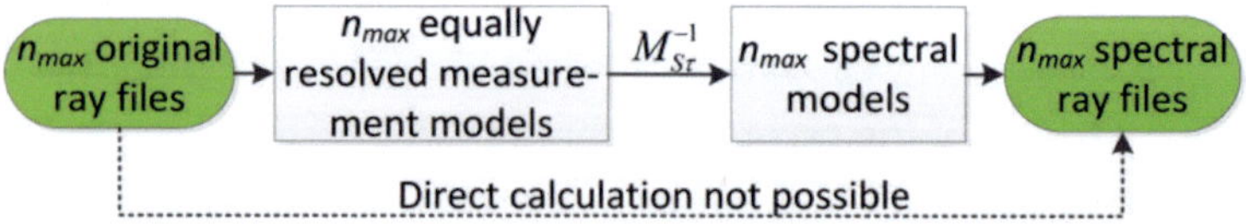

Figure 7.1: Concept to apply the reconstruction Eq. 4.17.

7.2 HISTOGRAM BASED MEASUREMENT MODELS

The model creation and calculation time should not exceed the overall measurement and creation time of the ray files because the calculation should not become the bottleneck of PMBS. Additionally, it would be advantageous, if the spectral model could be used directly for inverse ray tracing. This would require the models to be competitive with existing models in terms of memory requirements and access time. To achieve these aims, concepts from similar ray file based light source models [53], [54], [99] have been adapted.

7.2.1 FROM RAY FILE TO MODEL

The transformation of the original ray files to the models can be divided into three main steps, which are visualized in Fig. 7.2. The corresponding transformations for the case of ray files with constant amplitudes are illustrated in Fig. 7.3.

Step 1 focuses solely on the spatial information (X, Y, Z). Firstly, all rays are projected onto an enveloping spatial hemisphere to reduce the spatial dimension to two[2] and to enable the possibility of describing them with spherical coordinates (ϕ_S, ϑ_S) [99]. Subsequently, a grid defined by the resolution parameter N_{spatial} is created and all rays are assigned to a specific bin to create a 2-dimensional histogram. Figure 7.3 shows the generation of the spatial histogram on the left hand side. The overall model after step 1 is illustrated in Fig. 7.2 (a). The false color represents the measurement amplitudes, which are the number of rays.

[2] Typically the ray files already start on an envelope if they are generated with commercial measurement equipment.

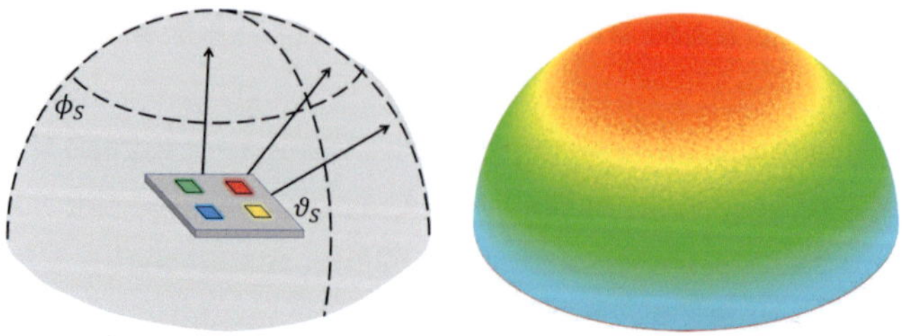

(a) Step 1

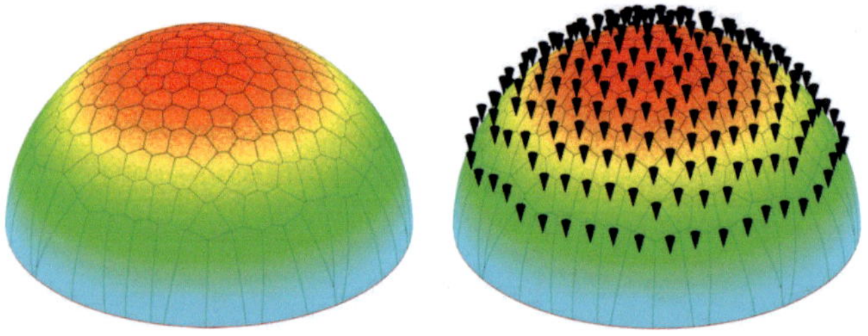

(b) Step 2 **(c)** Step 3

Figure 7.2: Visualization of overall model on enveloping hemisphere following the creation of the spatial histogram (step 1), the generation of the local areas (step 2) and the individual assignment of the angular distribution curves to their specific local area (step 3).

Step 2 defines different local regions within the spatial model as shown in Fig. 7.2 (b). The envelope is divided into N_{Voron} regions such that each region covers approximately the same amount of ray starting points as shown on the upper part of the right hand side in Fig. 7.3. That way it is ensured that each region has approximately the same amount of rays. As suggested in [54], this thesis uses vector quantification (more specific Lloyds algorithm [100]) to distribute the points associated to the voronoi regions for this step.

Step 3 is similar to step 1 but focuses solely on the angular ray directions (ϕ, ϑ), by creating one angular distribution curve (ADC) for each voronoi region. The ADCs are described as histograms with N_{angular} bins based on the directions of all rays in the voronoi region and can thus be interpreted as point source models. In total there are N_{Voron} ADCs. The process is visualized in the right hand side of Fig. 7.3 and the final model in Fig. 7.2 (c).

In this thesis, the spatial hemispheres as well as all ADCs are divided into Cartesian coordinates with approximately constant solid angles as described in [99], [101], [102]. First, the hemisphere is divided into constant polar angular steps. Subsequently, the azimuth is divided into constant angular steps depending on the current polar angle, such that a certain solid angle is defined. The resolution parameter N_{spatial} or N_{angular} determines the value of that solid angle. This ray allocation offers the possibility of classifying the rays using one-dimensional kd-search trees [99], which allow a very fast classification with a calculative complexity $O(\log N_{\text{Ray}})$. Firstly, a 1-dimensional search tree is used for the polar angles. The leaves of that tree consist of 1-dimensional kd-search trees for the azimuth angles. Alternative possibilities of allocating the rays are introduced and compared in [103].

While all resolutions N_{spatial} and N_{angular} and the positions and number of the voronoi regions N_{Voron} have to be the same for all models, it should be mentioned that it is advantageous for the memory requirements for the resolution of all ADCs to be the same. This offers the possibility of saving the angular positions only once.

Furthermore, the ADCs can be stored as sparse matrices because large parts of the ADCs equal zero since the local angular distributions on the enveloping hemisphere are quite directional. Although the concept works for both kind of ray files (constant and variable amplitudes), ray files with constant amplitudes are slightly advantageous as all ray weighting can be ignored. The generation of ray files from the model described in the next subsection also creates rays with constant amplitudes.

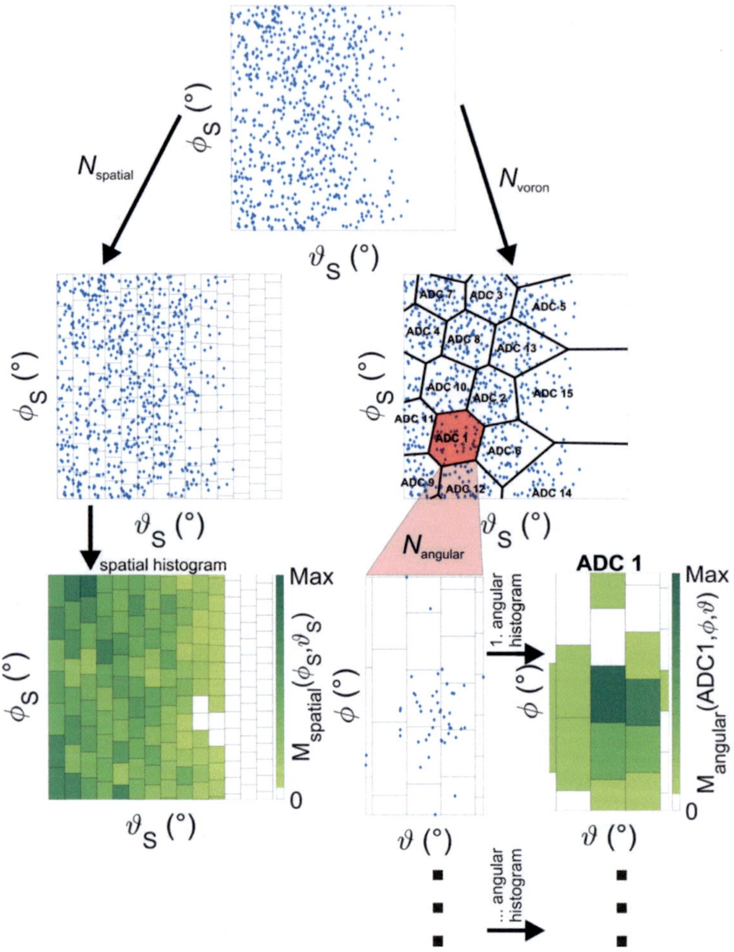

Figure 7.3: Visualization of model creation process: The left hand side shows the determination of the spatial histogram and the right hand side the generation of the angular model, which consists of N_{Voron} different local regions and their individual angular histograms.

7.2.2 FROM MODEL TO RAY FILE

Since a histogram $M_{\text{spatial}}(\phi_S, \vartheta_S)$ is an estimation of a probability function, the normalized spatial histogram can be interpreted as the probability of occurrence of a random ray starting point (X, Y, Z) on the hemisphere. That ray starting point also defines the voronoi region $Voron(\phi_S, \vartheta_S)$ and thus its specific angular histogram. The angular histograms $M_{\text{angular}}(Voron(\phi_S, \vartheta_S), \phi, \vartheta)$ can be normalized and interpreted similarly as the probability of occurrence of a random ray direction (ϕ, ϑ). Since there is an ADC for each voronoi region, $M_{\text{angular}}(Voron(\phi_S, \vartheta_S), \phi, \vartheta)$ also depends on the starting point. Finally, the overall model value (or if normalized, the probability of occurrence) of a ray is given by the product of $M_{\text{spatial}}(\phi_S, \vartheta_S)$ and $M_{\text{angular}}(Voron(\phi_S, \vartheta_S), \phi, \vartheta)$ as described in Eq. 7.1.

$$M(\phi_S, \vartheta_S, \phi, \vartheta) = M_{\text{spatial}}(\phi_S, \vartheta_S) \times M_{\text{angular}}(Voron(\phi_S, \vartheta_S), \phi, \vartheta) \tag{7.1}$$

The creation of a ray file from the model is basically the inverse of the model creation process. Firstly, randomized starting points are created. Then the starting points are assigned to a specific ADC. Finally, the randomized ray directions are assigned to the starting points. As suggested in [54], the random variables are created using inverse transform sampling of the spatial histogram $M_{\text{spatial}}(\phi_S, \vartheta_S)$ and the ADCs $M_{\text{angular}}(Voron(\phi_S, \vartheta_S), \phi, \vartheta)$ respectively. Furthermore, it is necessary to randomly distribute the starting points within each bin. This distribution has to take into account the original allocation process. In the case of Cartesian coordinates with approximately constant solid angles as described above, the randomized values in ϕ/ϕ_S direction can be distributed equally within the bin. In the polar direction the distribution is defined as sinusoidal because the solid angle (or surface area) is proportional to the sinus of its polar angle. This is important because an equal distribution leads to artifacts at low polar angles.

7.2.3 PRECISION ESTIMATION

The precision of the generated ray file is determined by three different resolution parameters N_{spatial}, N_{angular} and N_{Voron}. However, it also depends on the number of rays in the original ray file and the measurement resolution during the goniophotometric measurement. Furthermore, the precision of the spectral ray files cannot be estimated in general as there is no reference data. However, it is possible to estimate the precision of the quantification by comparing the original measured ray files to ray files, which were created based on the measurement models as shown in Fig. 7.4.

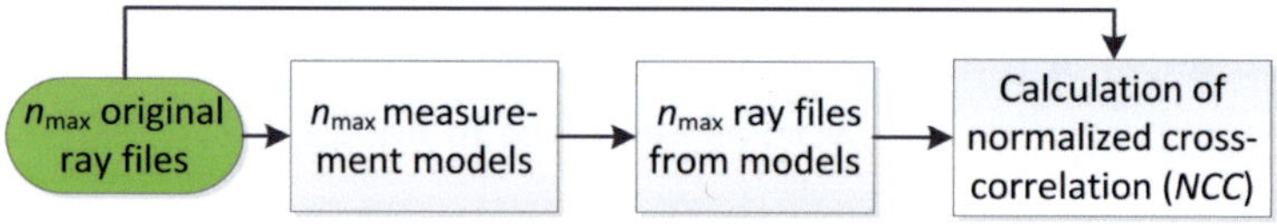

Figure 7.4: Validation principle of generated models and ray files.

As proposed in [68], [69] the similarity of two ray files is obtained by evaluating the normalized cross-correlation (NCC) values of their angular distributions at different distances r to cover near-, mid- and far field. The NCC of measurement model n at distance r is defined in Eq. 7.2.

$$NCC_{n,r} = \frac{\sum_{\phi,\vartheta} \left(I_{n,r}(\phi,\vartheta) - I_{\text{mean},n,r}\right) \times \left(J_{n,r}(\phi,\vartheta) - J_{\text{mean},n,r}\right)}{\sum_{\phi,\vartheta} \left(I_{n,r}(\phi,\vartheta) - I_{\text{mean},n,r}\right)^2 \sum_{\phi,\vartheta} \left(J_{n,r}(\phi,\vartheta) - J_{\text{mean},n,r}\right)^2} \tag{7.2}$$

The variable $I_{n,r}(\phi,\vartheta)$ is the achieved pseudo-intensity distribution[3] at distance r of a ray file, which was created with measurement model n, and the variable $J_{n,r}(\phi,\vartheta)$ is the achieved pseudo-intensity distribution

[3] Technically speaking, it is a near field pseudo-intensity because it refers to a finite distance r. The term intensity distribution is only defined in the case of $\lim_{r\to\infty} r$.

at distance r of the original ray file, which was used to create measurement model n. The variables $I_{\text{mean},n,r}$ and $J_{\text{mean},n,r}$ are the mean values of the corresponding pseudo-intensity distributions. Then the minimal value

$$NCC = \min_{n,r} NCC_{n,r}. \tag{7.3}$$

is used as the figure of merit for the precision of the measurement model.

Each pseudo-intensity distribution is achieved by assigning the rays to equally distributed points on an enveloping hemisphere as described above. Unfortunately, $I_{n,r}(\phi, \vartheta)$ and $J_{n,r}(\phi, \vartheta)$ also require a resolution parameter. However, this resolution can be achieved by obtaining the NCC between the "same" measurement ray file. This requires the ray file to be created twice using the same measurement raw data. If the generation parameters are changed slightly (for instance the number of rays, or the start value in the case of RIGO801) the ray file is only statistically identical to the first. Then these ray files can be compared at different resolutions. If the resolution gets too high, the NCC will drop below the suggested value of $NCC = 0.99$ [68], [69].

7.3 HISTOGRAM BASED SPECTRAL MODELS

All resolutions N_{spatial}, N_{Voron} and N_{angular} as well as all positions of the voronoi regions are the same for each measurement model. In that way the reconstruction Eq. 4.17 can be performed on the spatial histogram $M_{\text{spatial},n}(\phi_S, \vartheta_S)$ and $M_{\text{angular},n}(Voron(\phi_S, \vartheta_S), \phi, \vartheta)$ to create the histograms $A_{\text{spatial},n}(\phi_S, \vartheta_S)$ and $A_{\text{angular},n}(Voron(\phi_S, \vartheta_S), \phi, \vartheta)$ of all models. Finally, these histograms contain the amplitudes of the physical basis spectra and can be used to create a spectral ray file as described above.

7.3.1 NEGATIVE AND FALSE POSITIVE AMPLITUDES

A high *NCC* requires high resolution parameters but this also leads to a problem, namely the occurrence of negative reconstruction results in the spectral models. Figure 7.5 shows three measurement histograms $M_1(\phi, \vartheta) - M_3(\phi, \vartheta)$ and one resulting spectral histogram $A_1(\phi, \vartheta)$ following the application of the reconstruction Eq. 4.17 at different resolutions. If the resolution increases, the ray number in the bins further decreases until statistically empty bins start to arise in the measurement models (white bins in Fig. 7.5). This is caused by the discrete nature of the original ray files. In that case, bins are statistically subtracted from empty bins. This results in negative values (yellow to red bins in Fig. 7.5) at these bins, which indicate a mismatch between the measurement models. Higher resolutions lead to more statistically empty bins and a stronger occurrence of negative amplitudes.

As each $A_{\text{spatial},n}(\phi_S, \vartheta_S)$ and $A_{\text{angular},n}(Voron(\phi_S, \vartheta_S), \phi, \vartheta)$ represents an amplitude of a physical basis spectrum, negative amplitudes A_n^- can be assumed to be solely reconstruction errors. However, as the negative amplitudes occur due to a mismatch of a minuend with its supposed subtrahend, the non-subtracted minuends lead to false positive amplitudes. Even if the non-physical negative amplitudes are ignored (or eliminated during the inverse transform sampling), the false positive amplitudes lead to visible artifacts in the spectral models and the created spectral ray files. Therefore, the amount of negative/false positive amplitudes $Q_n(A_n)$, which is the relation of negative amplitudes $|A_n^-|$ to all amplitudes $|A_n^-| + A_n^+$, is defined as a quality metric for each spectral model

$$Q_n(A_n) = \frac{\sum |A_n^-|}{\sum |A_n^-| + \sum A_n^+} \tag{7.4}$$

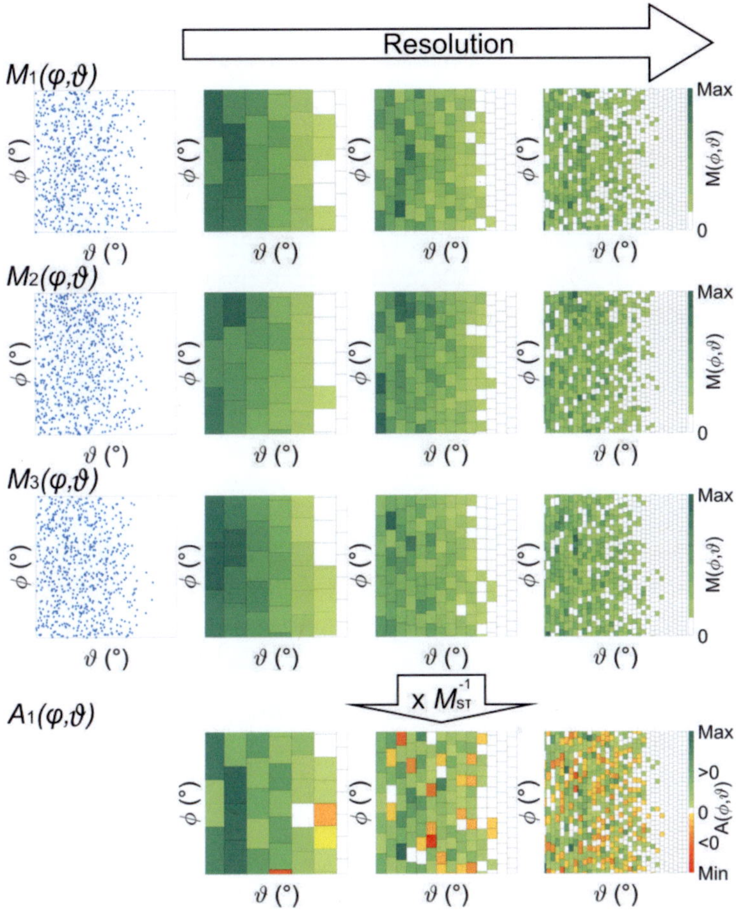

Figure 7.5: Occurrence of negative amplitudes in spectral histogram $A_1(\phi, \vartheta)$ caused by increasing mismatch between the different measurement models $M_1(\phi, \vartheta) - M_3(\phi, \vartheta)$ with increasing resolution.

with

$$A_n^- \in A_n \forall A_n < 0 \quad A_n^+ \in A_n \forall A_n \geq 0.$$

A low value of $Q_n(A_n)$ indicates that there are only a few negative/false positive amplitudes and therefore only few artifacts in the spectral model. Unfortunately, the resolutions required to obtain a low $Q_n(A_n)$ contradict the resolution requirements and lead to a low NCC. In the next subsection an algorithm which reduces $Q_n(A_n)$ without affecting the NCC is introduced.

7.3.2 NEAREST NEIGHBOR CALCULATION

The low amount of negative amplitudes in the case of low resolutions shows that there is no fundamental mismatch between the measurement models. Since the mismatch only arises in the case of high resolution models, it can be concluded that it arises solely at high frequencies due to the discrete nature of the ray files and that the false positive values therefore occur in the neighborhood of the negative values. While false positive values cannot directly be distinguished from correct positive values, all negative values can assumed to be artifacts as described above. This important physical boundary condition always holds in the case of PMBS since the reconstructed amplitudes are always those of the physical basis spectra. These relations are the basis of the algorithm displayed in Fig. 7.6. The regions which contain artifacts due to incomplete calculations are detected by searching for negative amplitudes. The calculation is then completed by compensating the positive values, which are likely to be false positive, with the negative values in the neighborhood region until either the negative or positive amplitudes vanish. The process repeats itself while increasing the searched neighborhood region until a user defined threshold $Q_{\text{final},n}$ is reached.

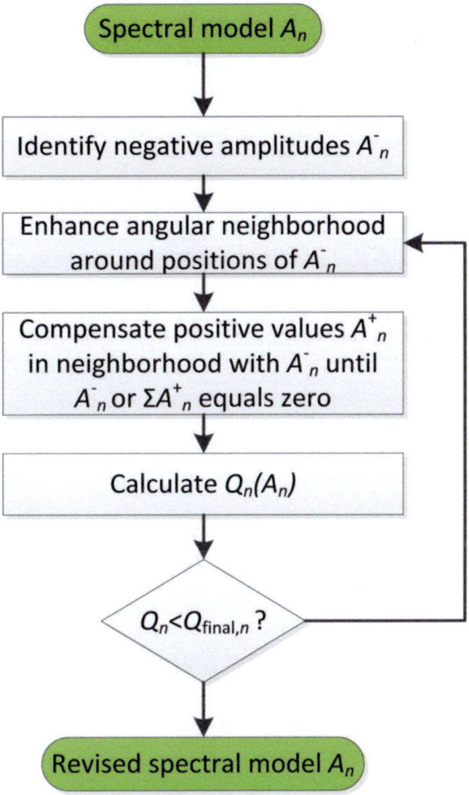

Figure 7.6: Workflow of nearest neighbor calculation.

The overall integral of the model remains constant during the process but $Q_n(A_n)$ reduces. The nearest neighbor calculation takes place in the spatial histogram and all ADCs. If the resolution of the models is very high and a large part was not calculated correctly, the nearest neighbor calculation performs the remaining calculation and therefore reduces both negative values and false positive values without affecting the resolution of the model.

7.4 CALCULATION PROCEDURE

Due to the improvement generated by the nearest neighbor calculation for models with high resolutions, the resolution can be optimized until a user defined NCC is reached. The auxiliary variable calculation success ζ is defined as the $\frac{1}{\zeta}$ fraction of incorrectly calculated bins in the spectral models. If the calculation success is 100, statistically only 1/100 of all bins are incorrectly calculated. If it is one, all bins were incorrectly calculated. The amount of negative amplitudes $Q_n(\zeta)$ of a spectral model directly depends on the calculation success and the matrix row of the reconstruction matrix used to create the spectral model. The relation is described in Eq. 7.5.

$$Q_n(\zeta) = \frac{\frac{|\alpha_n|}{\zeta}}{\beta_n - (|\alpha_n| - \frac{|\alpha_n|}{\zeta}) + \frac{|\alpha_n|}{\zeta}} = \frac{|\alpha_n|}{\zeta \times (\beta_n - |\alpha_n|) + 2\,|\alpha_n|} \quad \zeta \geq 1 \tag{7.5}$$

The variable α_n is the sum of all negative matrix values and the variable β_n is the sum of all positive matrix values occurring in the nth row of the reconstruction matrix (see Eq. 4.16 and 4.17). The calculation success, which only depends on the resolution parameters of the models, is used as an auxiliary variable to determine the histogram resolutions N_{spatial} and N_{angular}. It not only traces both resolutions back to one parameter, but also rates the interaction of the original ray files according to the reconstruction equation and allows a direct estimation of the initial calculation artifacts according to Eq. 7.5.

The final workflow is shown in Fig. 7.7. The user defines the desired precision as minimal NCC at different distances and $Q_{\text{final},n}$ for the neighborhood calculation. To start the optimization, an initial value for the calculation success ζ is required in addition. At first the $n_{\max}$ original ray files are transformed into $n_{\max}$ histogram based measurement models using arbitrary low initial model resolutions N_{spatial}, N_{Voron}

and N_{angular}. Subsequently, the $n_{\max}$ spectral models are calculated by applying the reconstruction Eq. 4.17. Equation 7.4 is used to estimate the current artifacts $Q_n(A_n) = Q_{\text{current},n}$ of the spectral models. The initial calculation success ζ defines a $Q_n(\zeta) = Q_{\text{aim},n}$ via Eq. 7.5. Then $Q_{\text{current},n}$ and $Q_{\text{aim},n}$ are compared. Algorithms such as bisection or false position are used [104] until $Q_{\text{current},n} \approx Q_{\text{aim},n}$ to adapt N_{spatial} for the spatial histogram and N_{angular} for all ADCs. These algorithms only require the assumption that the functional relation is one-dimensional and monotone and an allowed tolerance ϵ. Strictly speaking, ϵ is also an input parameter. However, if the precision of the false position or bisection algorithm is small, for instance below 1% of $Q_n(\zeta)$, its influence can be neglected.

If $Q_{\text{current},n} \approx Q_{\text{aim},n}$ for the most critical spectral model, $n_{\max}$ ray files are created from the measurement models to calculate their *NCC* values with the original ray files as described in the section above. The *NCC* obtained is compared with the desired *NCC*. If the obtained *NCC* is smaller than the desired *NCC*, the remaining resolution parameter N_{Voron} is optimized for the given calculation success ζ. If N_{Voron} is too small, the *NCC* in the far field gets higher than in the near field because few ADCs with a high angular resolution are associated with large starting areas. If N_{Voron} is too large, the *NCC* in the far field is smaller than in the near field for the same reason.

The model introduces a fuzziness between far field and near field or rather the angular distributions resolution and its associated area for a given calculation success ζ, which is adjusted with N_{Voron}. If the desired *NCC* cannot be reached by adapting N_{Voron}, the initial ζ is reduced. If ζ can remain high, the initial artifacts are small. However, if ζ becomes small, the spectral models show artifacts. The process stops when the desired *NCC* is obtained. Finally, the nearest neighbor calculation reduces the initial calculation artifacts and completes the creation of the hyperspectral LED models.

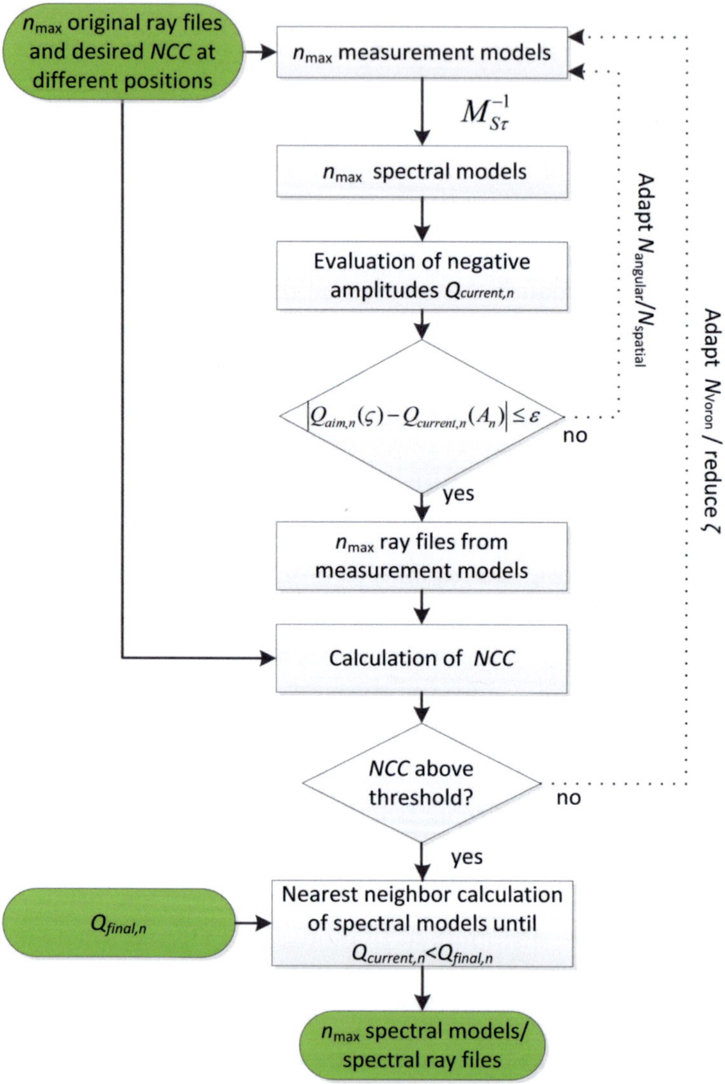

Figure 7.7: Workflow of the model creation and calculation process including the determination of the resolution parameters N_{spatial}, N_{Voron} and N_{angular} by using the precision parameters NCC and $Q_{\text{final},n}$.

7.5 VALIDATION OF CALCULATION PROCEDURE

The concept in Fig. 7.7 works for every arbitrary LED spectrum which might also contain phosphor conversion. Nevertheless, each step is illustrated by the example of the spectral reconstruction of the RGB LED F50360 from Seoul Semiconductor [105]. The RGB LED is chosen because the spectrum and position of each spectral source are well separated and therefore well known. Thus, the RGB LED enables the possibility of easily detecting spectral reconstruction errors and offers a reasonable validation of the whole calculation process described in this chapter.

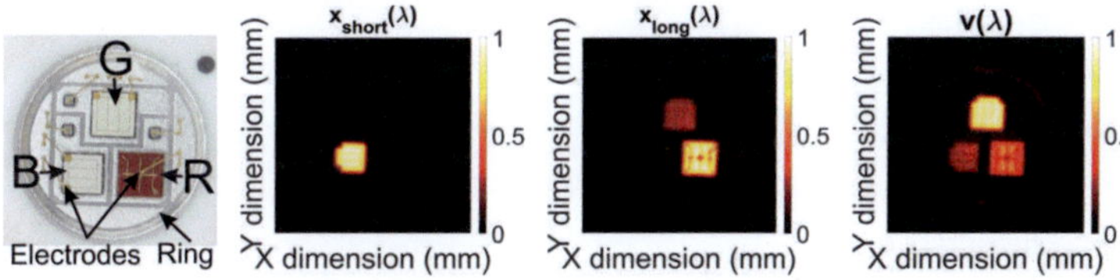

Figure 7.8: RGB LED: Photographic image and normalized pseudo-irradiances $M_1(X,Y) - M_3(X,Y)$ of the measured ray files in the LED plane.
$M_1(X,Y)$: $s_{\tau,1}(\lambda) = x_{\text{short}}(\lambda)$, $M_2(X,Y)$: $s_{\tau,2}(\lambda) = x_{\text{long}}(\lambda)$, $M_3(X,Y)$: $s_{\tau,3}(\lambda) = v(\lambda)$.

Figure 7.8 shows the test device and the normalized pseudo-irradiances $M_1(X,Y) - M_3(X,Y)$ of all measured ray files with 10×10^6 rays in the LED plane. This plane is chosen since the irradiance provides a sharp image of the individual LEDs, which can be verified by a comparison with the photographic image. The electrodes of the LED as well as a ring can be observed in the measured ray data[4]. The LED at the top of the images is the green LED. The blue LED is on the left hand side and the red LED on the right hand side.

[4] The ring may be hard to see in a printed version of this thesis.

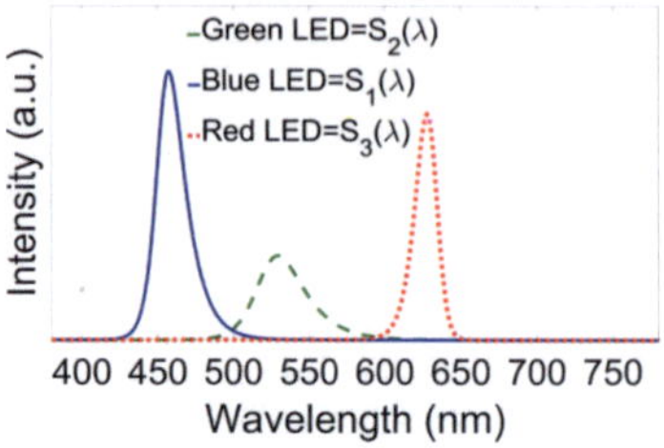

(a) Basis spectra $S_n(\lambda)$

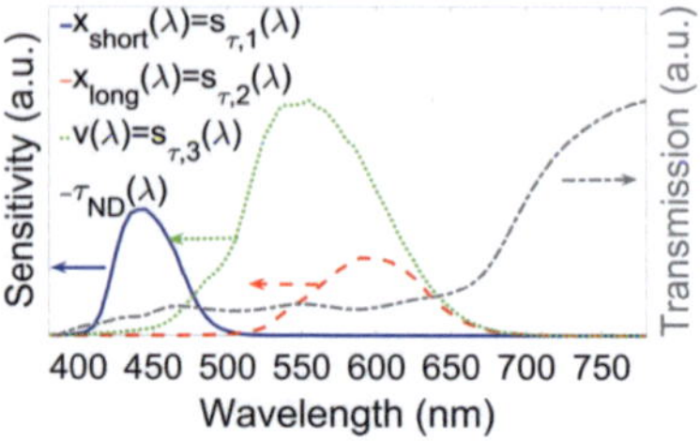

(b) Sensitivities $s_{\tau,\mathrm{eff},n}(\lambda)$

Figure 7.9: Input data for $M^{-1}_{\mathrm{S}\tau,\mathrm{RGB}}$: (a) Normalized basis spectra $S_n(\lambda)$ of well known RGB LED and (b) Normalized measured $s_{\tau,n}(\lambda)$ and $\tau_{\mathrm{ND}}(\lambda)$ used to create the ray files $M_1(X,Y) - M_3(X,Y)$.

The basis spectra of the test RGB LED are shown in Fig. 7.9 (a). The standard functions $x_{\mathrm{short}}(\lambda)$, $x_{\mathrm{long}}(\lambda)$ and $v(\lambda)$ and a neutral density filter as shown in Fig. 7.9 (b) were used to create these ray files. According to Eq. 4.16 and Eq. 4.17 the basis spectra $S_n(\lambda)$ and the effective spectral sensitivities $s_{\tau,\mathrm{eff},n}(\lambda)$ define the reconstruction matrix $M^{-1}_{\mathrm{S}\tau}$ shown in Eq. 7.6. The relative weightings of the measured ray files are incorporated by the normalization of the matrix $M_{\mathrm{S}\tau}$ as described in subsection 4.2.2.

$$M^{-1}_{\mathrm{S}\tau,\mathrm{RGB}} = \begin{bmatrix} 1.0051 & 0.0035 & -0.0086 \\ -0.2235 & -0.8428 & 2.0663 \\ 0.0564 & 1.4984 & -0.5548 \end{bmatrix} \qquad (7.6)$$

It should be noted that the filters used are not optimal according to chapter 6, but they provide a strong spectral overlap and therefore allow the validation of the calculation procedure. The first matrix row shows that the first measurement with $s_{\tau,\mathrm{eff},1}(\lambda) = x_{\mathrm{short}}(\lambda) \times \tau_{\mathrm{ND}}(\lambda)$ covers approximately one spectral source because the first matrix value is nearly one and the remaining values are orders of magnitude smaller. The first row describes the reconstruction of the blue LED. However, the reconstruction of the green and red LED requires a linear combination of all measured ray files, because neither $s_{\tau,\text{eff},2}(\lambda)$ nor $s_{\tau,\text{eff},3}(\lambda)$ measured only the individual green or red LED. The second filter $s_{\tau,\text{eff},2}(\lambda) = x_{\text{long}}(\lambda) \times \tau_{\text{ND}}(\lambda)$ measured a linear combination of the green and red LED and the third filter $s_{\tau,\text{eff},3}(\lambda) = v(\lambda) \times \tau_{\text{ND}}(\lambda)$ measured a linear combination of all LEDs. This is also indicated by the second and third row of $M^{-1}_{\text{S}\tau,\text{RGB}}$, because at least two values in each row are in the same order of magnitude.

Before the calculation procedure can start, the resolution and positions of the NCC calculation and the desired NCC, as well the $Q_{\text{final},n}$ need to be defined. As the envelope of all measured ray files was set to 5 mm, the intensity distributions were evaluated to calculate the NCC (see subsection 7.2.3) at the distances 6, 7, 8, 9, 10, 20, 30, 40, 50 and 300 mm. Each of the equally distributed sampling points used to determine the intensity of the RGB LED covered approximately a solid angle of 0.0025 sr. At this resolution the NCC between the original ray files containing 10^6 rays and the original ray files containing 10×10^6 rays from the same measurement is above 0.996 at each distance. The difference can assumed to be noise in the smaller ray file resulting from the high resolution. As an NCC of above 0.99 indicates good agreement [68], [69], that resolution is suitable for the precision estimation of the created light source models.

Figure 7.10 visualizes the irradiance in the LED plane of the reconstructed green LED spectral model for different desired NCCs (without nearest neighbor calculation). The amount of negative amplitudes gets larger with increasing resolution parameters. If $Q_n(A_n)$ is as small as 1% or 10%, only the green LED is represented, but the irradiance is also blurred out due to the low angular and spatial resolution of the histograms, which is quantitatively verified by the NCC. However, if the resolutions are very high, spectral artifacts as high as 26% or 33%, which are caused by false positive amplitudes, start to occur. The model assigns parts of the blue and the red LED also to the green LED.

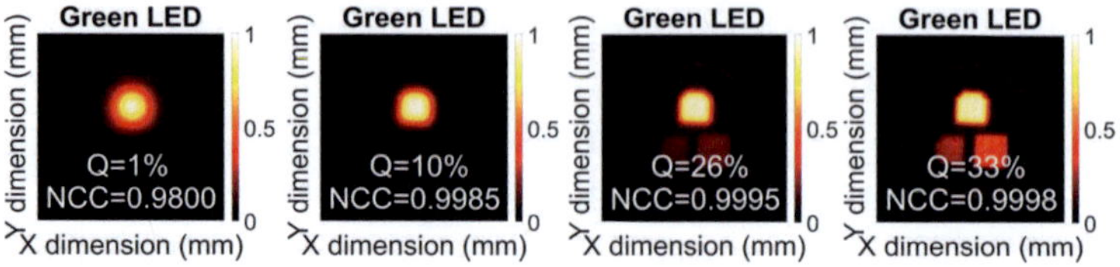

Figure 7.10: Trade off between model precision NCC and artifacts as amount of negative/false positive amplitudes $Q_2(A_2)$ (green LED model) as normalized irradiance in the LED plane.

The relation between this incomplete calculation and the NCC is also visualized in Fig. 7.11. The graphs show the achieved NCC displayed as $1 - NCC$ on a logarithmic scale at the tested distances for different calculation successes ζ with original ray files containing 10^6 rays in Fig. 7.11 (a) and 10×10^6 rays in Fig. 7.11 (b). The NCC with the same calculation success is always higher in Fig. 7.11 (b). This is to be expected, because higher model resolutions can be achieved, before the noise of the larger and therefore higher resolved original ray files reduces the calculation success. The graphs also verify that the calculation success ζ remains high for NCCs above 0.99 with typical original ray amounts, for instance 10^6 or 10×10^6. According to [68], [69] an $NCC > 0.99$, or rather $1 - NCC < 1 \times 10^{-2}$, indicates good conformity between model and measurement.

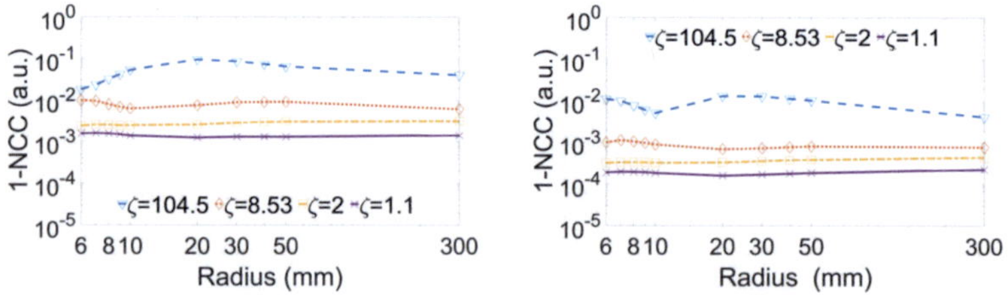

(a) 10^6 rays **(b)** 10×10^6 rays

Figure 7.11: Relation between calculation success ζ and the NCC with original ray files containing (a) 10^6 rays (b) 10×10^6 rays.

Figure 7.10 supports this suggestion as it shows the irradiance of the spectral models in the LED plane although the calculated NCC displayed in Fig. 7.10 refers to near field pseudo-intensity distributions of the measurement models at different distances as described in subsection 7.2.3.

Figure 7.12 visualizes the improvement due to the nearest neighbor calculation compared to the original calculation with a $NCC = 0.9998$ and $Q_{\text{final},n}=Q_{\text{start},n}/1000$. The variable $Q_{\text{start},n}$ refers to the amount of negative amplitudes of each spectral model before performing the nearest neighbor calculation. The artifacts caused by false positive rays are strongly reduced for both the green and red LED. All three LEDs can be seen. The comparison of Fig. 7.12 and Fig. 7.8 shows that the workflow is capable of creating spatially and angularly resolved spectral ray files.

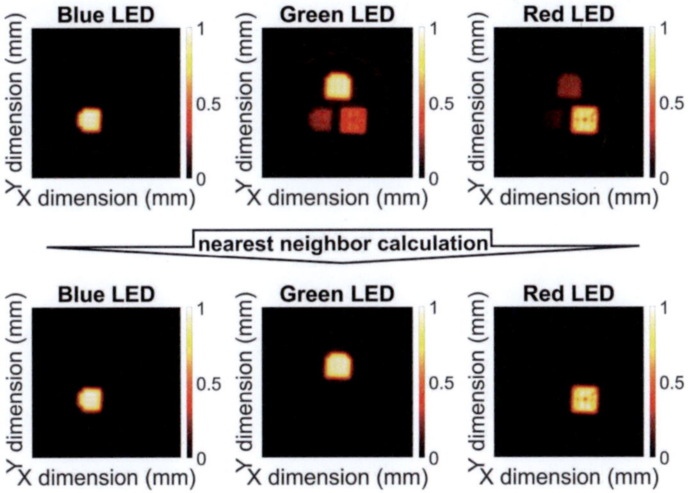

Figure 7.12: Normalized irradiance in the LED plane for all spectral LED models before and after the nearest neighbor calculation for $NCC = 0.9998$ and $Q_{\text{final},n}=Q_{\text{start},n}/1000$ for all models.

As the desired NCC of the measurement model was set to $1 - NCC < 3 \times 10^{-4}$, the reconstructed individual LEDs in Fig. 7.12 match the device shown in Fig. 7.8 with a high level of detail. Electrodes as well as reflections created by the surrounding ring are represented in the irradiance distributions created by the spectral models. Since the blue LED model is approximately represented by the first measurement ($x_{\text{short}}(\lambda)$), the reflections of the ring can be assumed to be represented well. The same reflection scheme is observed in the irradiance distributions of the green and red LED. The reflection is strong near the LED and nearly zero if the other LEDs shadow the ring.

Depending on the desired precision and using ray files with 10^6 rays, the simulation time of a MATLAB implementation was between 40 minutes on an average personal computer (Intel(R) Core(TM) i5-3470 CPU @ 3.2GHz and 16GB RAM) for an NCC of 0.996 and three hours on a simulation computer (Intel(R) Xeon(R) CPU E5-2670 v3 @2.3 Ghz and 98GB RAM) for the high precision $NCC > 0.9998$. Generating ray files using the histogram based models requires only a few seconds. Therefore the calculation time is the same order of magnitude as the measurement time.

CHAPTER 8

APPLICATION AND VALIDATION

In contrast to the examples and validations used in the previous chapters, the validations in this chapter use the complete workflow of PMBS to create spectral ray files. The basis spectra are always estimated as described in chapter 5. Subsequently, the goniophotometric measurements and calculations are performed. The first section applies PMBS on a phosphor converted white LED and compares the obtained results with data according to the Blue/Yellow approach. More complex spectral combinations are reconstructed in section 8.2. The last section verifies the spectral ray files obtained from section 8.2 by comparing ray tracing simulations of a simple optical system with far field measurements of the same system and the same LED source.

8.1 COMPARISON TO THE BLUE/YELLOW APPROACH

As described in section 3.4.4 and 4.1, the Blue/Yellow approach is a feasible, fast and therefore industrially used measurement method to create spectral ray files of phosphor converted white LEDs with two ILMD-based measurements. The main idea, which is the allocation of the physical sources to basis spectra and ray files, is identical to PMBS. While the Blue/Yellow approach is less complicated to apply, PMBS is a physically more valid approach and does not introduce a systematic error by virtue of its underlying assumptions. The required measurement amount of both methods is identical in the case of a phosphor converted white LED.

To quantitatively compare both methods, an LED with a strong angular color uniformity was chosen. Angularly resolved spectral measurements have been performed to assess the results. Figure 8.1 shows the LED as well as its color shadow on a photographic image. The angularly resolved spectral measurements are evaluated as chromaticity coordinates in the CIE $u'v'$ diagram and shown in Fig. 8.1 as well. As the color shift lies on an almost straight line, it can be concluded that the color shift is caused by the linear combination of the blue LED and the yellow phosphor. Therefore, both the Blue/Yellow and the PMBS approach can be used to create a spectral ray file of this LED.

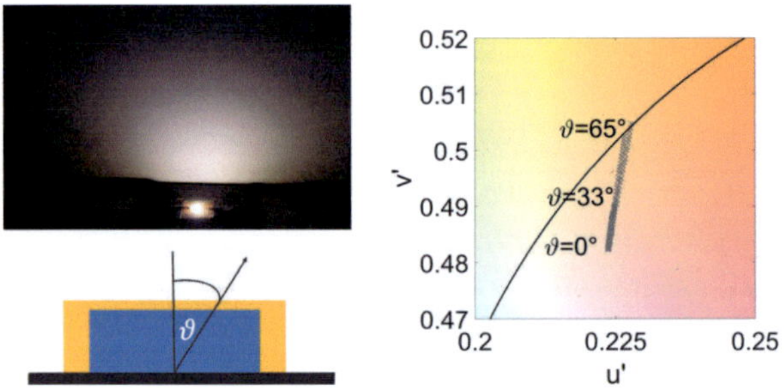

Figure 8.1: LED used for comparison and chromaticity coordinates obtained from angularly resolved spectral measurement.

The Blue/Yellow ray file of the LED was obtained from an industrial partner according to the state-of-the-art procedure[1]. The separation wavelength was set to 520 nm[2]. Figure 8.2 shows the sharp spectra and its theoretical color mixing line in addition to the measured chromaticities. The chromaticity shift is well described by the Blue/Yellow approach. There is only a small misalignment between the measured data and the theoretical color mixing line, which is caused by the

[1] It is important to note that the same LED was used for both Blue/Yellow and PMBS.

[2] It can be assumed that the separation wavelength was defined by the physical filters used to measure the blue and yellow ray files.

simplification of the sharp spectral separation as described in section 3.4.4 and 4.1.

The reconstructed spectra deviate stronger, the stronger the spectrum to be reconstructed deviates from the global spectrum used for the sharp spectral separation. As the global spectrum used for the separation is based on the sum of all measured spectra, the largest deviations occur at the main radiance direction $\vartheta = 0°$ and the largest polar angle measured at $\vartheta = 65°$. They are displayed in Fig. 8.3 and not only show a characteristic discontinuity at the separation wavelength, but also differ stronger in their relation and their relative distribution in the region of the spectral overlap.

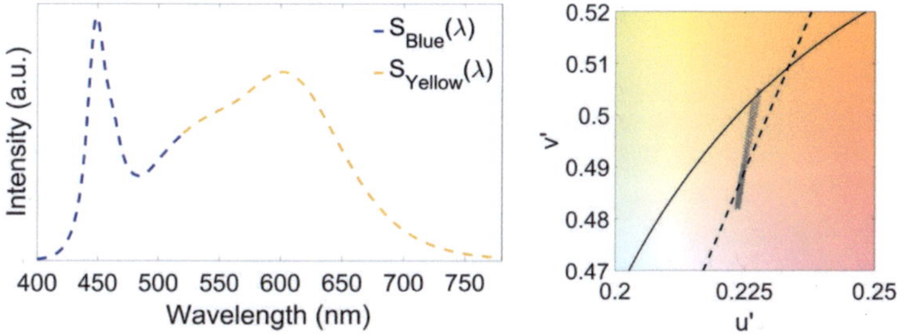

Figure 8.2: Sharp separated spectra and associated mixing line of Blue/Yellow approach compared to angularly resolved spectral measurements.

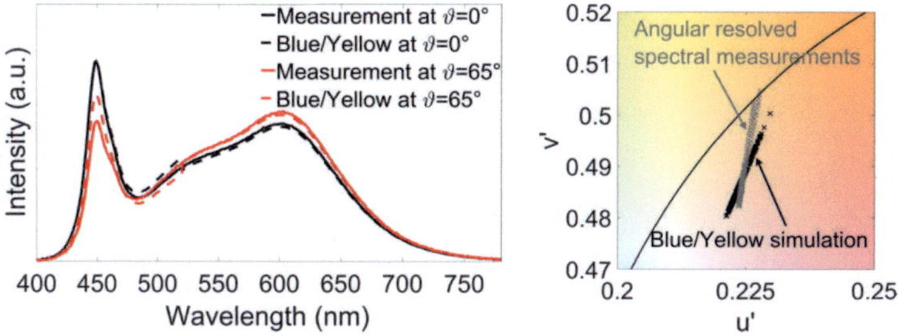

Figure 8.3: Comparison of angularly resolved spectral measurements to Blue/Yellow simulation of phosphor converted white LED.

The obtained chromaticities are also shown in Fig. 8.3. As indicated by the color mixing line, there is a misalignment between the measured chromaticities and the simulation results of Blue/Yellow. Furthermore, there seems to be a slight blue shift. But it can also be noted that the relative chromaticity shift of the LED is well presented by the Blue/Yellow approach. It should be mentioned that this state-of-the-art method can be improved by using a shorter separation wavelength. However, the separation wavelength always remains an arbitrary parameter.

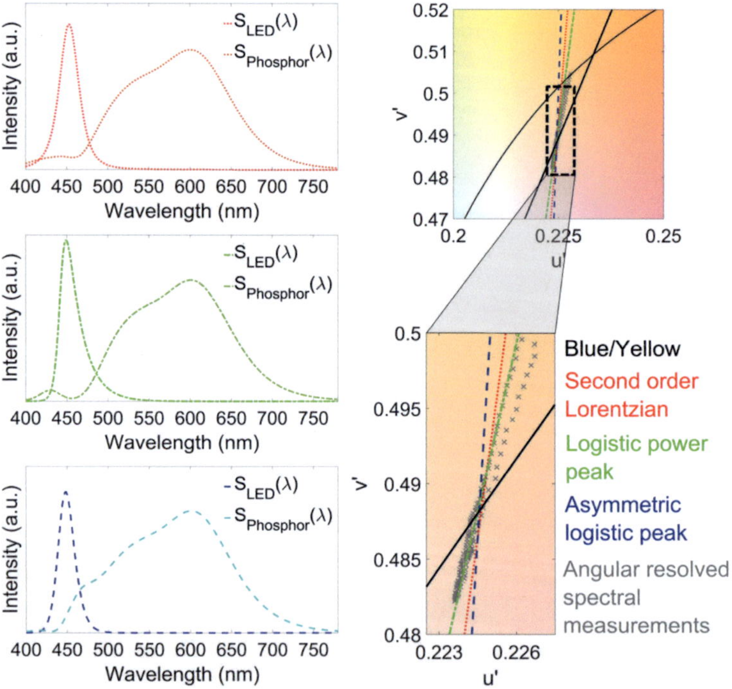

Figure 8.4: Basis spectra obtained from initial modeling and their associated color mixing line as well as Blue/Yellow color mixing line.

The basis spectra estimation of PMBS is also based on the global spectrum used for the sharp spectral separation. Figure 8.4 shows estimated basis spectra models and their theoretical color mixing lines. The alignment of the color mixing lines compared to the angularly resolved spectral measurements is used to select the final basis spectra. As the color mixing line of the basis spectrum Logistic power peak (green dashed-dotted line) shows the best alignment, it is chosen for PMBS. Furthermore, it should be noted that all PMBS basis spectra are more suited to describe the measurement values compared to the sharp spectral separation (solid line). The ILMD-based measurements have been performed with the standard optical filters for $x_{\text{short}}(\lambda)$ and $x_{\text{long}}(\lambda)$ and an additional neutral density filter. It is an advantage if the reconstruction works well with these filters because they are often available in ILMDs, which are able to measure colorimetric values. Their suitability is to be expected because the system is simple and the filters separate the basis spectra well. The camera luminous fluxes are used to obtain the absolute weighting of the ray files. Finally, the reconstruction is performed.

The results of the PMBS simulation are visualized in Fig. 8.5. The comparison of reconstructed spectra to measured spectra in Fig. 8.5 shows that the deviations are small in the case of PMBS. The spectra differ more strongly in the spectral region of the basis spectra overlap but the overall differences are smaller than those of the Blue/Yellow spectra in Fig. 8.3.

The comparison of the reconstructed chromaticities to measurements in Fig. 8.5 and the comparison to Fig. 8.3 lead to the same conclusion. PMBS is not only able to describe the relative chromaticity shift more precisely, but also differs less in terms of the absolute chromaticity differences, whose mean value is $\Delta u'v'_{\text{mean}} \approx 0.002$. Note that these results are achieved although the obtained basis spectra still differ from the true basis spectra.

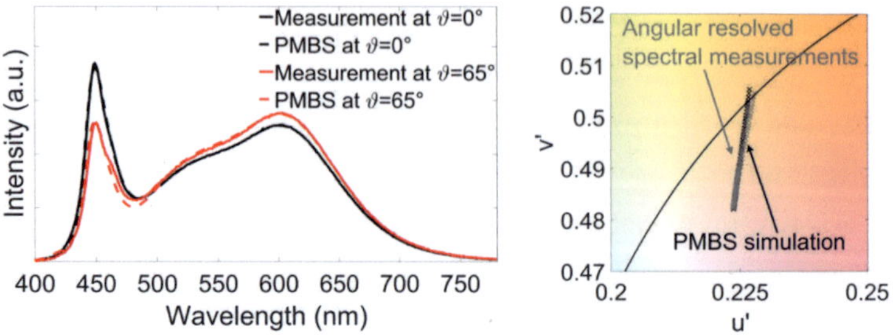

Figure 8.5: Comparison of angularly resolved spectral measurements to PMBS simulation of phosphor converted white LED.

The small intensity peak of the phosphor between 400 nm and 420 nm as well as the right hand side of the LED basis spectrum from the Logistic power peak can be assumed to be artifacts. However, as in the case of the Blue/Yellow approach, the combination of both weighted basis spectra lead to a cancellation of the individual basis spectra deviations and thus to a good representation of the true LED spectra.

8.2 RECONSTRUCTION OF TYPICAL SPECTRAL LED COMBINATIONS

Although PMBS performs well compared to the Blue/Yellow approach in the case of phosphor converted white LED, its main advantage is the generality for more complex LED systems. This section shows reconstruction results of two more complex but typical LED systems. In both cases, the measurement object shown in Fig. 8.6 from [106] is used as an example. An aperture set creates a strong angular and spatial color variation in the near field, limits the solid angle and thus reduces the measurement time of this object [106]. This temperature controlled device was operated in two different colorimetric setups, namely Red/White and Red/Green/Blue/White. Both setups represent typical cases of modern lighting technology such as color tunable luminaries.

The following two subsections each focus on one specific example and can be divided into four different parts. The first part is the initial basis spectra estimation. The second part focuses on the measurements and a first precision estimation. The third part describes the basis spectra optimization and the last part assesses the final results.

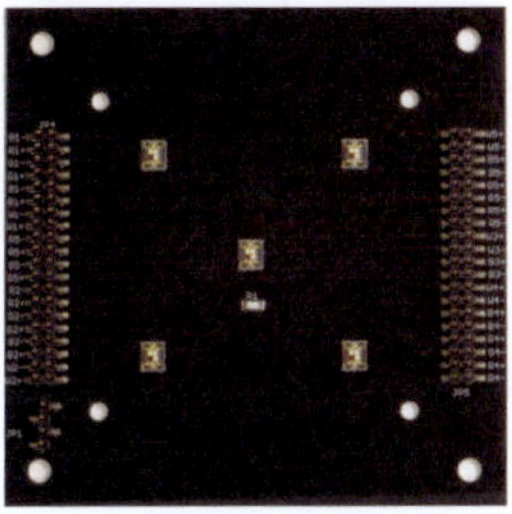

(a) Board

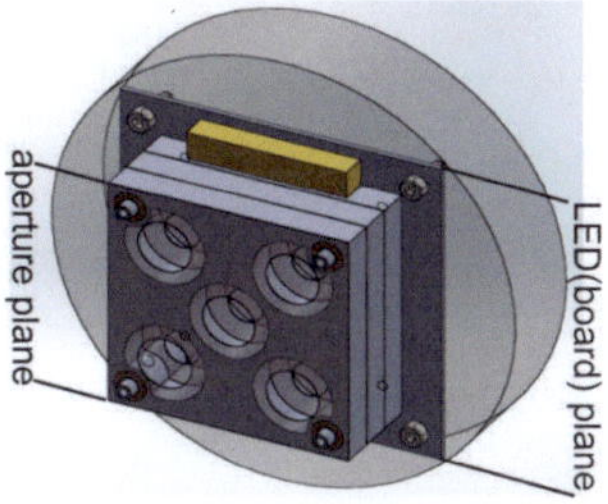

(b) Schematic sketch

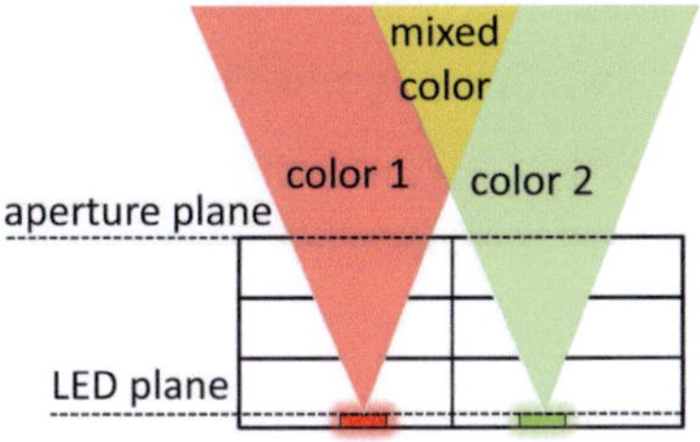

(c) Mixing principle

Figure 8.6: Adaptive test device with apertures introduced in [106] (images adapted from [106], photo (a) from Tino Weiss, LTI).

Initially, all measurements are performed with standard optical filters to test PMBS with typically available filters. The filter selection according to chapter 6 is only applied, if the standard filters cannot reconstruct the spectra. Generally, the filters shown in Fig. 8.7 are considered. Interference filters are not used as they are more prone to deviations of the initial modeled basis spectra as shown in section 6.4. The relative weighting of the measured ray files are always based on the camera luminous flux values because the inverse reconstruction assumes correct basis spectra, which cannot be ensured.

The desired normalized cross-correlation (NCC) in the calculation procedure was set to 0.9997. The NCC of the intensity distributions was evaluated at 1.1, 1.2, 1.5, 2, 2.5, 5 and 10 times the radius of the enveloping hemisphere as well as at infinity.

During the reconstruction the spatial separation is assessed in the LED (board) plane. The plane is marked in Fig. 8.6 (b). The LED plane offers the highest localization and is thus automatically chosen during the spatial separation algorithm of the basis spectra assessment and during the basis spectra optimization. Only this plane ensures the evaluation of the spatial separation between the individual chips within one RGBW LED.

However, it is hard to visualize artifacts and reconstructions in this plane. Therefore, the reconstructions and measured ray files are all shown in the aperture plane, which is also visualized in Fig. 8.6 (b) and (c).

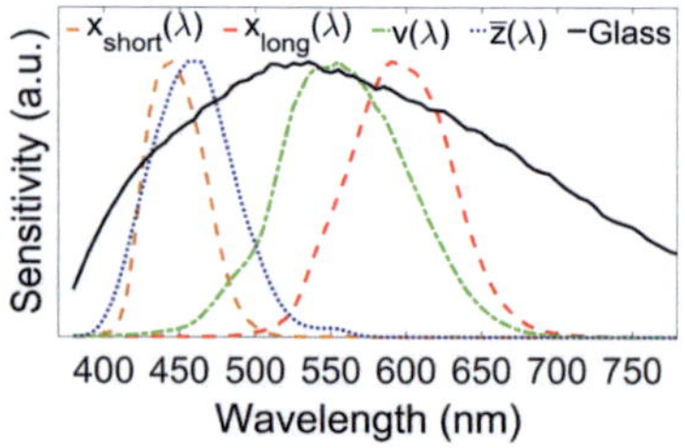

(a) Standard filters

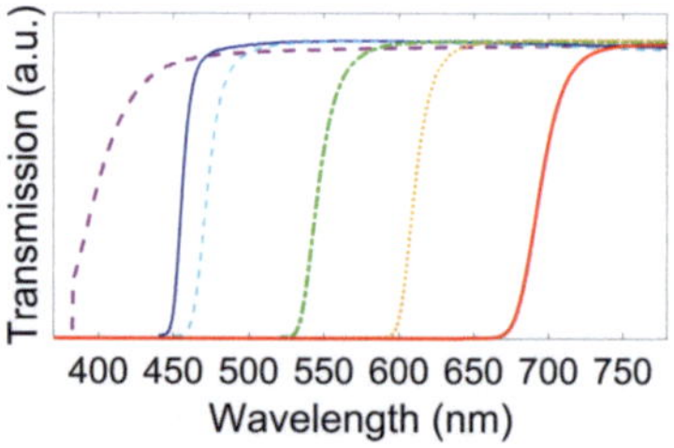

(b) Glass absorption filters

Figure 8.7: Filters considered for the spectral reconstructions of Red/White and Red/Green/Blue/White.

8.2.1 RED/WHITE (RW)

The first setup combines white (W) and red (R) LEDs, which is a typical combination to enhance colorimetric parameters as for instance the color rendering index. The specific setup and angularly resolved spectral measurements are shown in Fig. 8.8. Due to the apertures, there is a strong angular and spatial chromaticity shift in the near field. The colors mix partly with an increasing distance from the measurement object. The chromaticity shift is nearly linear because the blue LED and the yellow phosphor ratio does not change significantly within the aperture limited solid angles. Nevertheless, the spectral sources are modeled individually as red and blue LED as well as a phosphor.

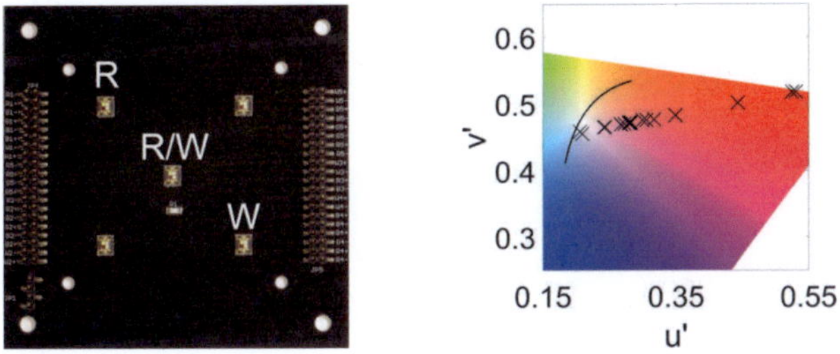

Figure 8.8: Configuration of Red/White (images adapted from [106]) and measured chromaticities CIE $u'v'$ diagram.

First, the initial modeling is used to obtain the three basis spectra from a spectrum measured at main radiance direction. The comparison of the obtained basis spectra to the individually measured spectra is shown in Fig. 8.9. The obtained spectra match the true basis spectra. The chromaticity distances of the individual LEDs are smaller than $\triangle u'v' = 0.001$. Only the estimated phosphor deviates slightly which leads to a chromaticity distance of $\triangle u'v' = 0.007$.

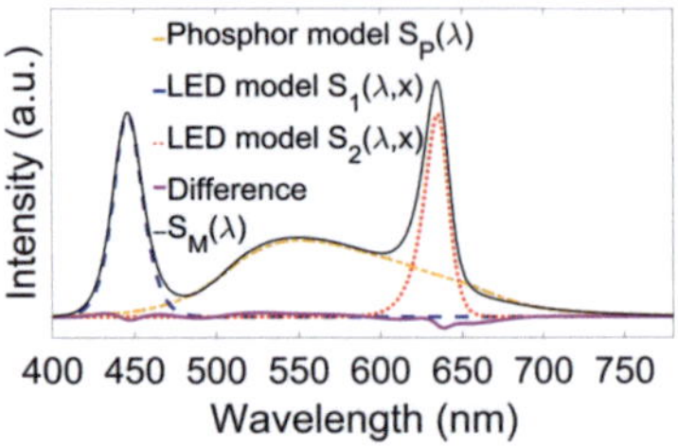

(a) Complete result

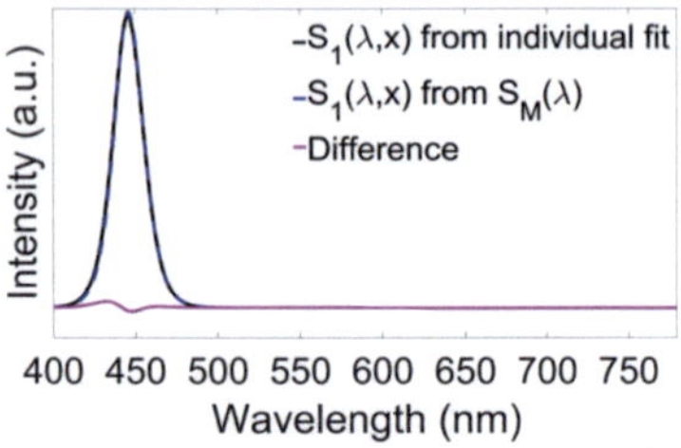

(b) Blue LED

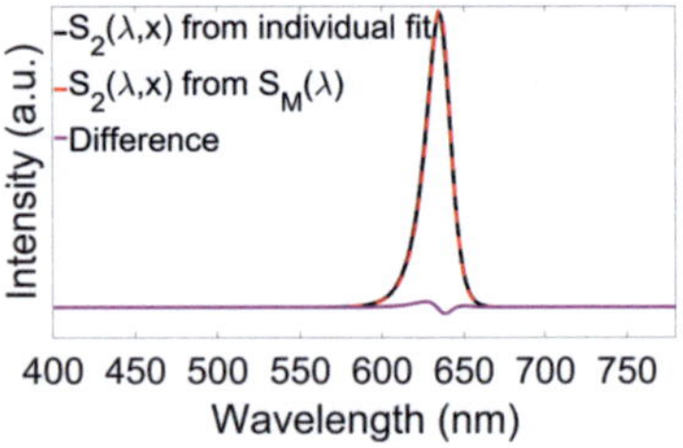

(c) Red LED

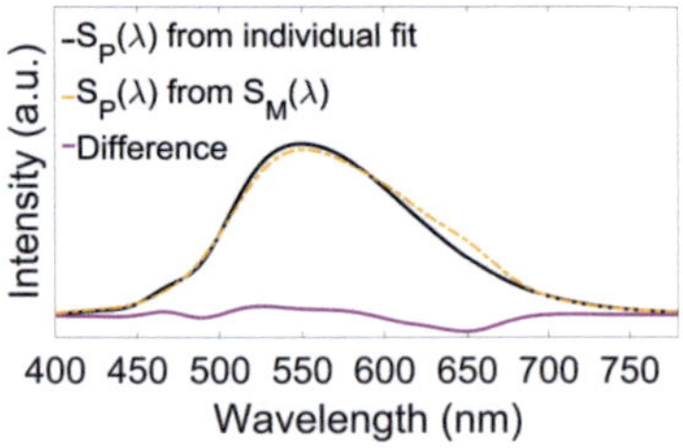

(d) Phosphor

Figure 8.9: Comparison of initial modeled basis spectra modeling to individual fitted basis spectra of Red/White.

Then the ILMD-based goniophotometric measurements using the standard filters for $x_{\text{short}}(\lambda)$, $x_{\text{long}}(\lambda)$ and $v(\lambda)$ are performed. Figure 8.10 (a) shows the obtained measurement ray files as normalized pseudo-irradiances in the aperture plane. The shapes are not circular and differ for the individual measurements. This non-circular shape occurs since the individual LED chips are off centered. The apertures are centered on the virtual midpoint of the RGBW LED.

The first estimation of the reconstruction result is shown as normalized irradiance in the aperture plane in Fig. 8.10 (b). The reconstructions each show two spectral sources and no significant artifacts apart from a slight amount of negative values in the red LED model. Therefore the measurements based on the standard filters are used to proceed with the basis spectra optimization using spatial separation.

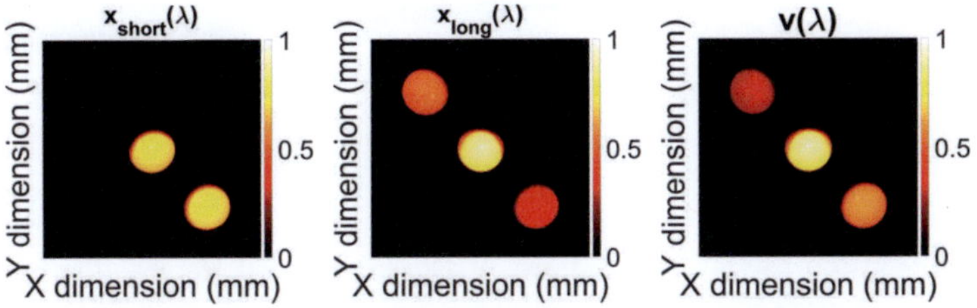

(a) Measured ray files

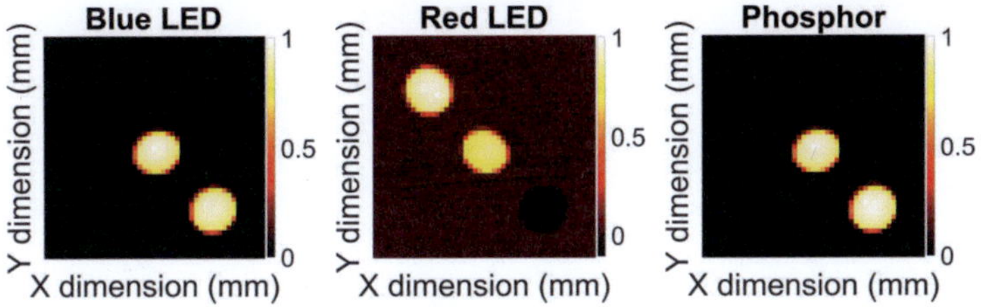

(b) Spatial separation

Figure 8.10: Ray files used for the Red/White reconstruction visualized as normalized pseudo-irradiances and obtained spatial separation as normalized irradiance in the aperture plane.

The optimized basis spectra are shown in Fig. 8.11. As the initial modeling already achieved good results, the basis spectra optimization does not change the model spectra significantly. This can also be seen by comparing the spatial separation using the basis spectra from the initial modeling in Fig. 8.10 (b) to the spatial separation using the optimized basis spectra in Fig. 8.12. The spatial separation obtained with the optimized basis spectra shows a slight improvement in the case of the red LED model. The chromaticity distance of the optimized phosphor spectrum supports this impression as it reduces to $\triangle u'v' = 0.004$.

Finally, the spectral ray files are reconstructed using the calculation procedure described in section 7.4. Figure 8.13 shows spectral ray files

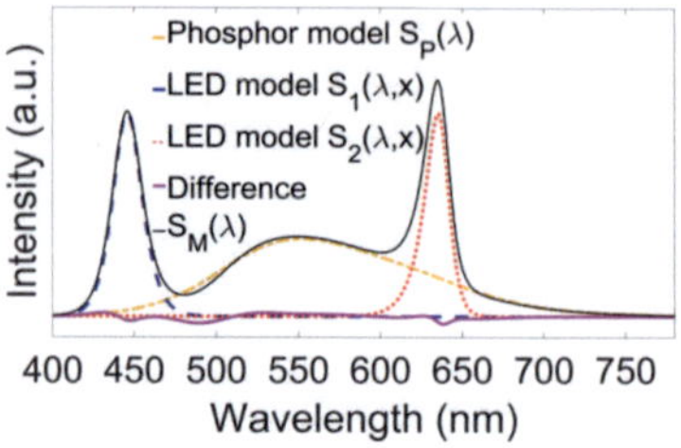

(a) Complete result

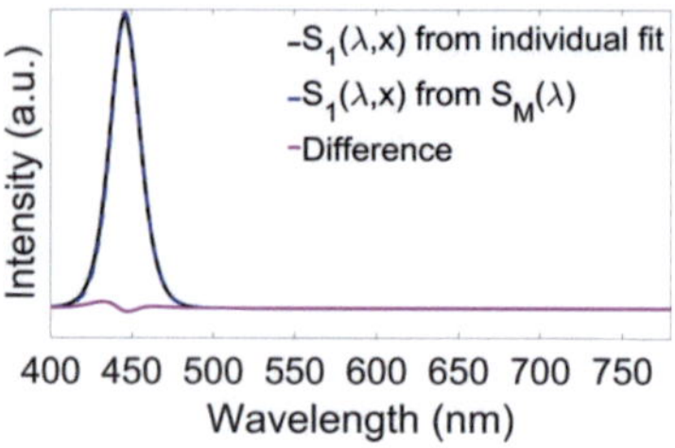

(b) Blue LED

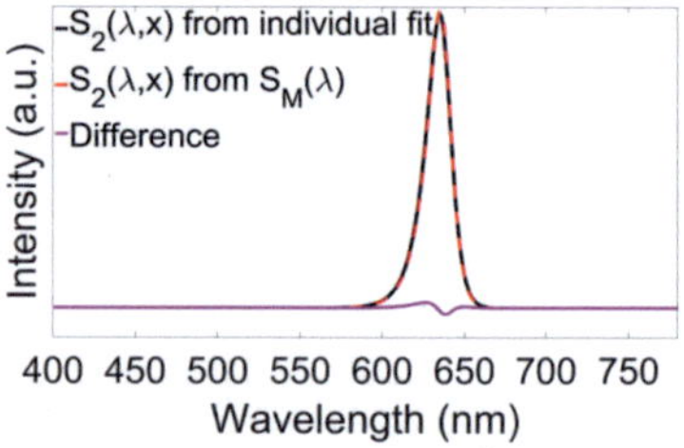

(c) Red LED

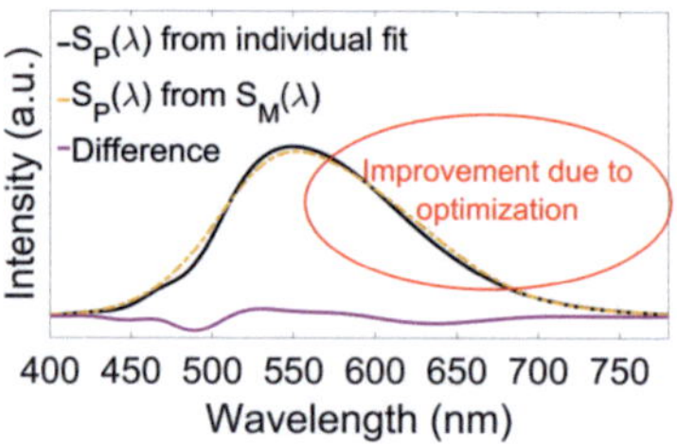

(d) Phosphor

Figure 8.11: Comparison of optimized basis spectra using spatial separation to individually fitted basis spectra of Red/White.

as normalized irradiance in the aperture plane as well as the chromaticity distance between the angularly resolved spectral measurements and spectral reconstructions on the goniometric measurement sphere of the spectroradiometer. The results obtained represent the setup defined in Fig. 8.8.

The irradiance plots of the phosphor and the red LED show a small artifact, which assigns a part of the red LEDs to the phosphor and vice versa. However, the wrongly assigned spectral parts are below 0.3 percent. In conformity with the measured ray files, the apertures are not circular and differ for the individual spectral models. The irradiances of the phosphors and the blue LEDs have the same shape because they are both part of the same white LEDs.

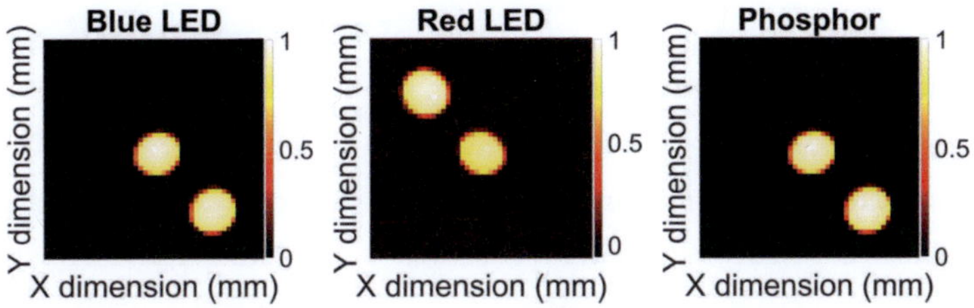

Figure 8.12: Spatial separation obtained with optimized basis spectra as normalized irradiance in the aperture plane.

The mean chromaticity distance of the spectral reconstruction on the virtual spectroradiometer sphere is $\Delta u'v'_{\text{mean}} \approx 0.005$ and therefore slightly higher than the median chromaticity distance $\Delta u'v'_{\text{median}} \approx 0.003$. These values also indicate that the spectral reconstruction performs well, even if only standard optical filters are used. The performance in an actual ray tracer is evaluated in section 8.3.

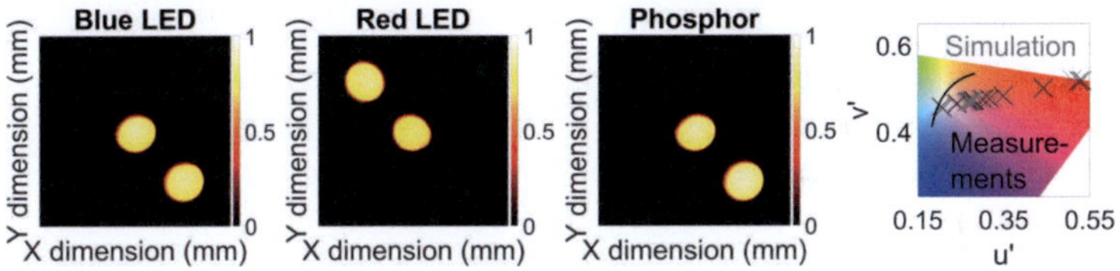

Figure 8.13: Obtained spectral ray files and reconstructed chromaticities compared to angularly resolved spectral measurements for the Red/White setup.

8.2.2 RED/GREEN/BLUE/WHITE (RGBW)

The second case, shown in Fig. 8.14, combines white (W), blue (B), red (R) and green (G) LEDs. RGBW LEDs are typically used in modern adaptive and intelligent lighting systems. The angularly measured chromaticities shown in the CIE $u'v'$ diagram include pure colors as well as mixtures. Therefore it is necessary to model all four basis spectra. Note that the blue LED chip of the white LED is assumed to be identical to the pure blue LED.

The initial basis spectra modeling was performed on a spectrum that had been measured at main radiance direction. Figure 8.15 shows the estimated basis spectra as well as a comparison to individually measured and modeled spectra. The chromaticity distance between the LED models is smaller than $\Delta u'v' \approx 0.003$. However, the phosphor basis spectrum shows a larger difference and a chromaticity distance as high as $\Delta u'v' \approx 0.037$.

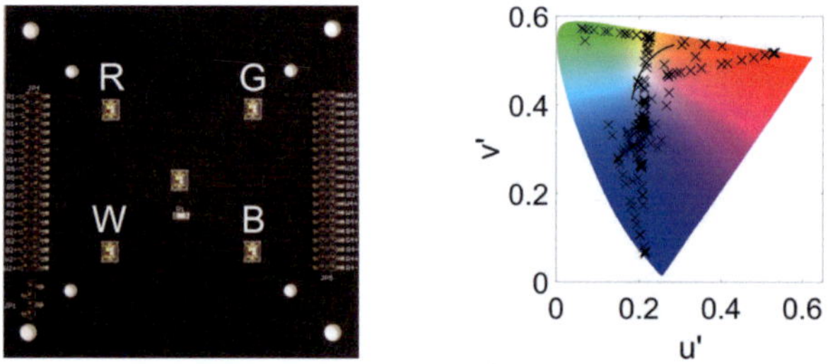

Figure 8.14: Configuration of Red/Green/Blue/White (image adapted from [106]) and measured chromaticity coordinates in the CIE $u'v'$ diagram.

The first measurements are based only on standard filters for $x_{\text{short}}(\lambda)$, $x_{\text{long}}(\lambda)$, $v(\lambda)$ and "Glass", which has no mentionable absorption. As in the case of Red/White and the phosphor converted LED, the filter selection was omitted such that in the first try only standard filters are used. The expected reconstruction performance is assessed using spatial separation, which is visualized in Fig. 8.16. It shows a weak performance. Instead of a spatial separation most LED models are located at the same positions. The LED in the lower left is assumed to be a green, red and blue at once. Furthermore, the spatial separation concludes that each LED has an additional phosphor.

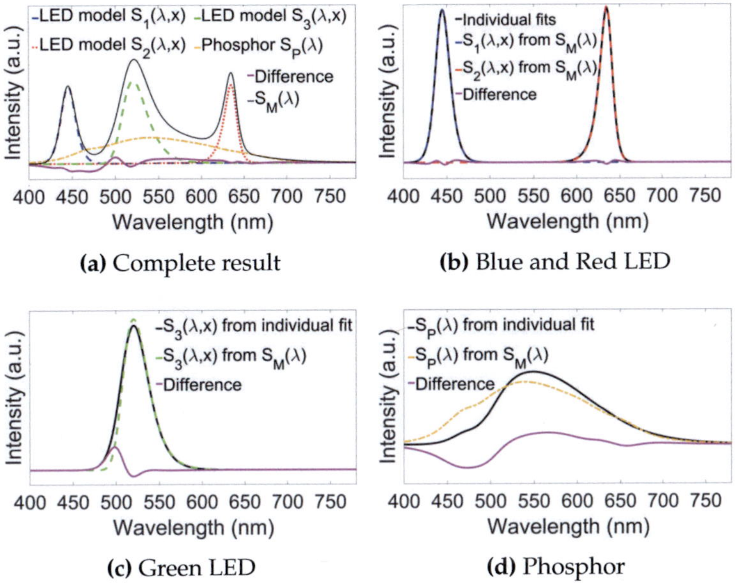

Figure 8.15: Comparison of initial modeled basis spectra modeling to individual fitted basis spectra of Red/Green/Blue/White.

This indicates a poor spatial separation. The weak performance may be caused by the deviations of the phosphor model, or by the chosen standard filters. Therefore, the estimated basis spectra shown in Fig. 8.15 are used to estimate a better suited filter set. The small inhomogeneity in the lower left corner is caused by spatial distortions due to an non-optimal lens calibration.

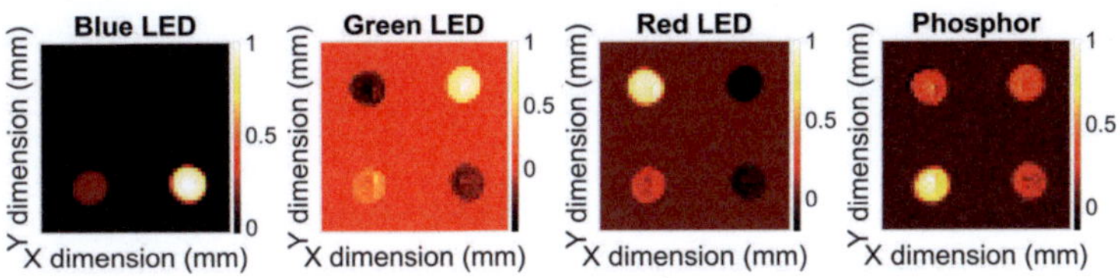

Figure 8.16: Spatial separation using standard optical filters and basis spectra after initial modeling.

The optimized measurements are performed with glass edge absorption filters with the cut off wavelengths 475, 610 and 695 nm and the standard filter for $x_{\mathrm{short}}(\lambda)$. The obtained measurement ray files as normalized pseudo-irradiances in the aperture plane are shown in Fig. 8.17. Again spatial separation is used to validate the reconstruction and to rate the basis spectra. The spatial separation in Fig. 8.18 still shows slight artifacts. There is a negative green spectral part at the location of the white LED. But overall the result is an improvement compared to the reconstruction based on standard filters in Fig. 8.16. The images show fewer LED locations, and reasonable combinations, as for instance a weak blue peak, which is located at a large phosphor peak and spatially separated colored LEDs.

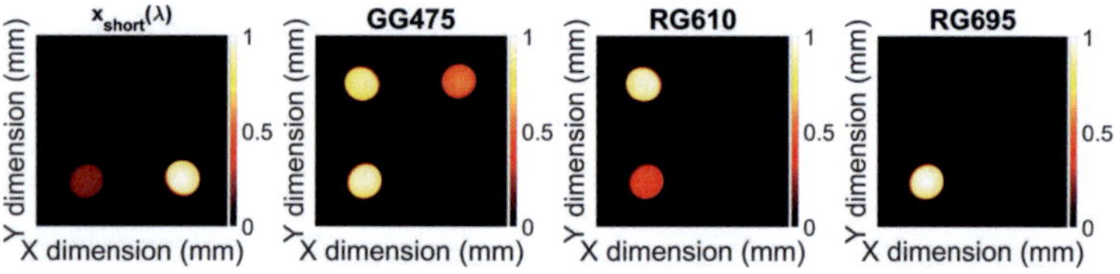

Figure 8.17: Ray files used for the RGBW reconstruction visualized as normalized pseudo-irradiances in the aperture plane.

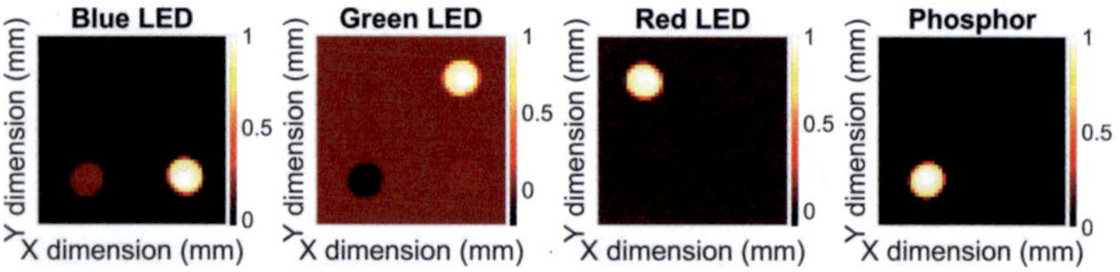

Figure 8.18: Spatial separation before basis spectra optimization using filter set according to filter selection procedure.

The basis spectra have been optimized using spatial separation. They are shown in Fig. 8.19 (a)-(d). While the LED spectra show no significant change regarding the colorimetric distance, the phosphor spectrum has improved and the chromaticity distance decreased to

$\Delta u'v' \approx 0.021$. The improvement is also verified by the spatial separation in Fig. 8.19 (e). In particular the green LED model shows fewer artifacts. However, there is still a negative spectral part from the red LED at the location of the white LED, which is an artifact. This artifact has to be considered in the calculation procedure according to chapter 7 by adapting $Q_{\text{final},n}$ to the negative values of the spatial separation. Otherwise the calculation converges slowly because it falsely assumes that all negative amplitudes are caused by the model resolutions.

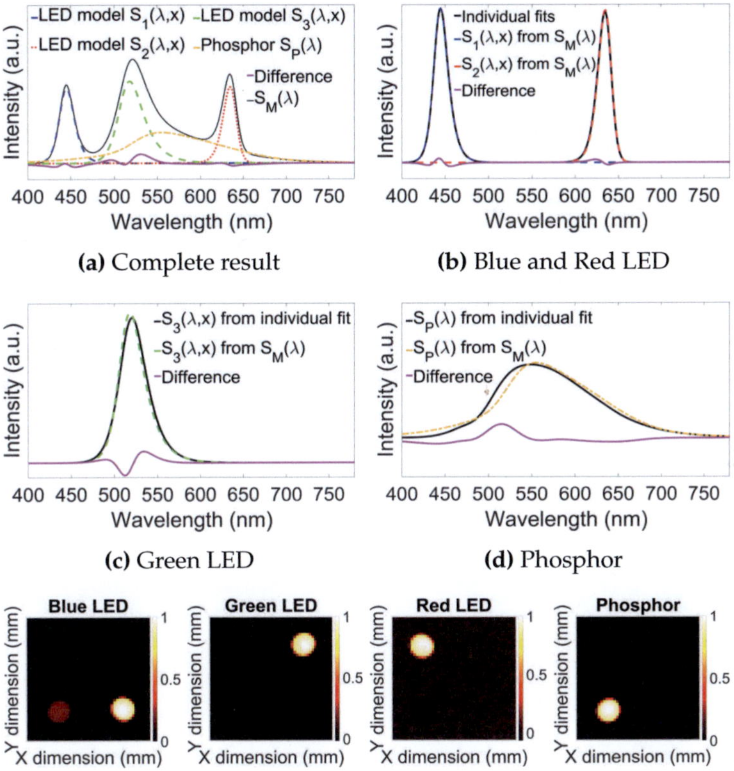

(a) Complete result

(b) Blue and Red LED

(c) Green LED

(d) Phosphor

(e) Spatial separation

Figure 8.19: Comparison of optimized basis spectra using spatial separation to individually fitted basis spectra of Red/Green/Blue/White.

The spectral models obtained are shown in Fig. 8.19. As the blue LED model is part of both the blue and white LED, the irradiance in the aperture plane shows a good spatial separation and thus represents the chosen setup. The green LED model shows an artifact, as it associates 5 percent of its radiometric power to that of the white LED. For the remaining models this value is below 1.5 percent.

The mean chromaticity distance of the spectral reconstruction on the spectroradiometer sphere is $\Delta u'v'_{\text{mean}} \approx 0.012$ and therefore higher than in previous examples. This value is mainly caused by comparably high deviations at high polar angles in the color mixing regions, which occur due to a decreased ray file resolution in these regions. Figure 8.20 shows a histogram of the chromaticity distances. Note that the median value, which ignores the outliers, provides a chromaticity distance of approximately $\Delta u'v'_{\text{median}} \approx 0.007$. The performance of the ray files in a ray tracer will be evaluated in the next section.

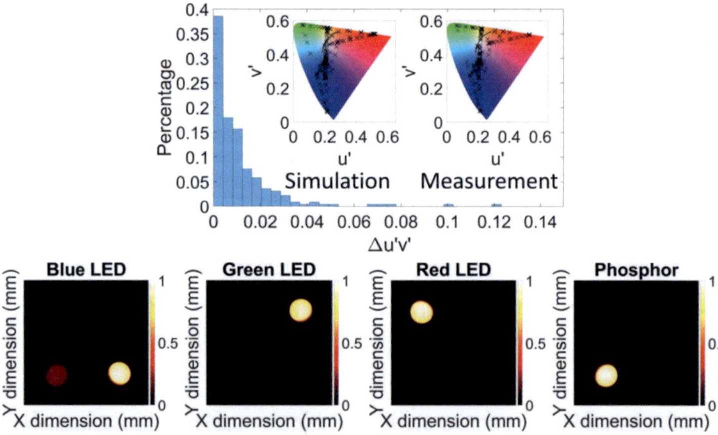

Figure 8.20: Obtained spectral ray files and reconstructed chromaticities compared to angularly resolved spectral measurements.

8.3 SPECTRAL RAY TRACING

To validate the suitability of the obtained spectral ray files, a far field measurement of a wavelength dependent optical system including the measurement object was compared to ray tracing simulations in the commercial ray tracer OptisWorks from OPTIS. The optical system consists of one dispersion prism because it shows a strong wavelength dependence and because the simple geometry minimizes deviations between the optical system and the CAD geometry, which can disturb the assessment of the spectral ray files.

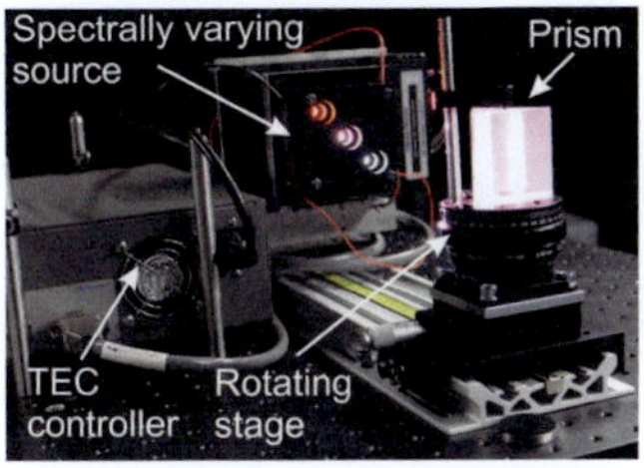

(a) Experimental setup (Red/White)

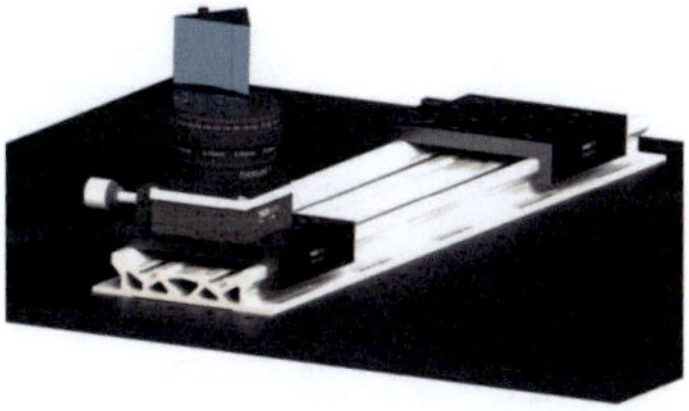

(b) Setup in simulation

Figure 8.21: Photographic image of experimental setup (a) and rendering of simulation setup (b) for the spectral ray tracing validation.

The experimental setup containing the temperature control (TEC controller) and the source as well as the dispersion prism on a rotating stage is shown in Fig. 8.21 (a). A colorimetric intensity distribution of the refracted part of the light was goniometrically measured with an automotive goniometer and the colorimeter C3300 from LMT.

The simulation setup is shown in Fig. 8.21 (b). In both simulation and measurement the prism was aligned with the rotating plate.

As in the measurement, the photometric center of the intensity sensor in the simulation was defined at the center of the prism. The ray file editor from OptisWorks was used to assign the basis spectra to the reconstructed ray files from section 8.2. Finally, a direct spectral simulation with the combination of all spectral ray files was performed. Each source was represented by 10×10^6 rays. A spectral intensity detector obtained the complete spectrum at each solid angle element.

As the prism was located in the near field of the source, the influence of the prism is different for each LED source. Figure 8.22 shows photographic images of the obtained distributions. Both distributions show the individual LEDs used in the setup.

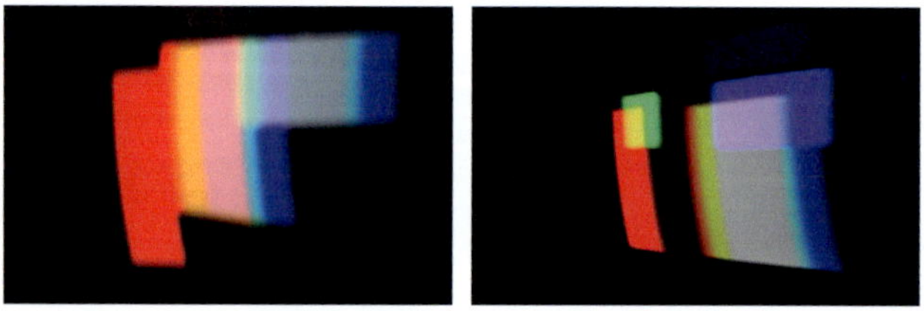

(a) Red/White **(b)** RGBW

Figure 8.22: Photographic images of obtained distributions.

The Red/White distribution consists of a solely white part in the upper right, a red/white mixture in the middle and a solely red part in the lower left. Furthermore, both parts with a white LED show a chromatic shift from blue to red caused by the prism. The characteristic shape arises due to the placement of the source and the aperture of the prism.

The RGBW distribution has similar characteristics. Each LED is located in a specific corner with an overlapping region. Both photographic images in Fig. 8.22 show only a part of the complete distribution. In the following subsections both examples are qualitatively and quantitatively evaluated to validate the obtained spectral ray files.

8.3.1 RED/WHITE (RW)

Figure 8.23 (a), (b) and (c) visualize measured as well as non-filtered and filtered simulated distributions of the Red/White setup. The filtering was performed with the OPTIS XMP-filter in order to decrease the noise in the simulation. Furthermore, only the three major sequences, which represent 92% of the radiometric power reaching the sensor, contribute to the simulation result[3]. The images aim to visualize both results in true colors. Therefore, values from the measurement and the simulation provided in X_{CIE}, Y_{CIE}, and Z_{CIE} were converted in RGB values. These RGB values were used to create the images. However, it has to be noted that the colors are still false colors if they are observed on a printed page or a display.

Nevertheless, the comparison of Fig. 8.23 (a) and Fig. 8.23 (b) shows a very good qualitative agreement between the simulation and measurement results of the Red/White source although the simulation still shows noise in the white regions. Both the general shape and color of each region as well as the overlapping regions obtained with the spectral ray files fit the measurement. This can also be verified by the red spot in the upper left corner. It is caused by a reflection of the red LED within the prism and occurs in both measurement and simulation. Only the curvature of the distributions seems slightly stronger in the measurement. It is likely that this difference is caused by a misalignment, which occurred either in the near field measurement or the simulation. Appendix B shows that the slight misalignment between measurement and simulation is not caused by the spectral ray files created with PMBS. It also shows that it is valid to consider only the three major sequences.

[3] The different sequences present different ray paths in the optical simulation and are typically used to suppress or analyse stray light. The sequence analysis was performed with the stray light analysis from OptisWorks.

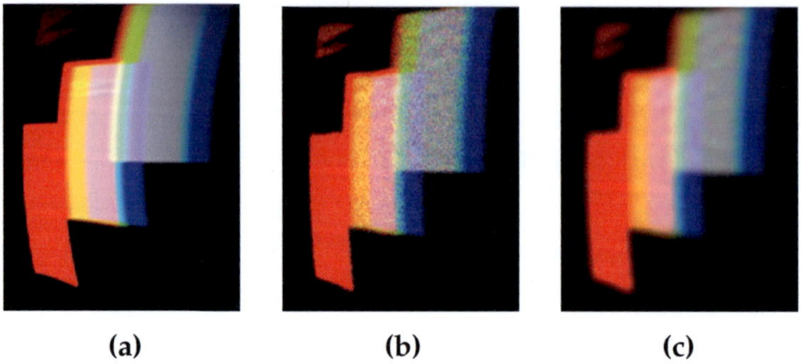

Figure 8.23: Comparison of far field measurement (a) and unfiltered (b) as well as filtered (c) spectral ray tracing with Red/White spectral ray file.

The quantitative comparison is shown in Fig. 8.24. Sub figure (a) visualizes the chromaticity distance between the chromaticity coordinates of the measurement and the chromaticity coordinates obtained from the spectral intensity sensor in the ray tracer. Note that only measurement and simulation values with a reasonable measurement signal are evaluated. The displayed colorimetric distance is limited to $\Delta u'v' \approx 0.05$ in order to provide a better overlook. Figure 8.24 (b) shows a histogram of the deviations and the mean chromaticity of the highlighted regions R, W and R/W. There are high deviations at the edges of the distributions, especially in the white region. It is likely that the main part of these deviations is caused by the slight misalignment between the distributions because the shape of the chromaticity distances above $\Delta u'v' \approx 0.05$ in Fig. 8.24 (a) is similar to the curvature in Fig. 8.23 (a).

Other causes of these deviations may be the applied XMP-filter, which smears the edges, or the number of rays used in the simulation. The phosphor spectrum serves as the probability density function for the phosphor ray file. Therefore, the number of rays associated with very short or very large wavelengths is comparably small. Furthermore, there are fewer rays in the solely white distribution. Both can be verified by comparing the noise in Fig. 8.23 (b) within the white and within the red/white distribution as well as at the chromatic edges of the white and red/white distribution. In both cases, the solely white region has more noise.

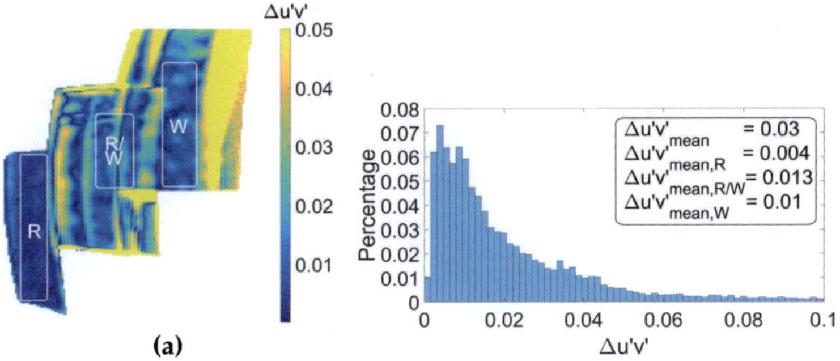

(a) Pixel-wise

(b) Distribution

Figure 8.24: Chromaticity distance $\Delta u'v'$ of measurement and spectral ray tracing with Red/White ray file.

Due to the deviations at the edges, the mean chromaticity distance is not suitable for quantifying the performance of the spectral ray files. The highlighted regions R, R/W and W in Fig. 8.24 (a) are less influenced by the misalignment and thus used. The deviations in the red region are as high as the basis spectra model precision from the used Asymmetric logistic peak (see Table 5.2 in chapter 5). The deviations in the white and red/white regions are higher but not solely noise induced. Note that the absolute deviations are within the color discrimination thresholds of white light colors in applications according to [62]. The main part of these deviations occurs due to an offset of $\Delta v' \approx 0.009$, which may originate from the phosphor basis spectra precision or the relative weighting between the phosphor spectrum and the blue LED or the uncertainty of the array spectroradiometer and the colorimeter used in the near or far field measurement. However, as the main deviation is induced by an offset, the relative deviations are expected to be smaller and enable for instance a uniformity analysis of the light distribution.

8.3.2 RED/GREEN/BLUE/WHITE (RGBW)

Figure 8.25 (a), (b) and (c) visualize measured as well as non-filtered and filtered simulated values of the RGBW distribution. As in the Red/White simulation, only the three major sequences, which represent 92% of the radiometric power reaching the sensor, contribute to the simulation result. Again the images aim to visualize the results in true colors.

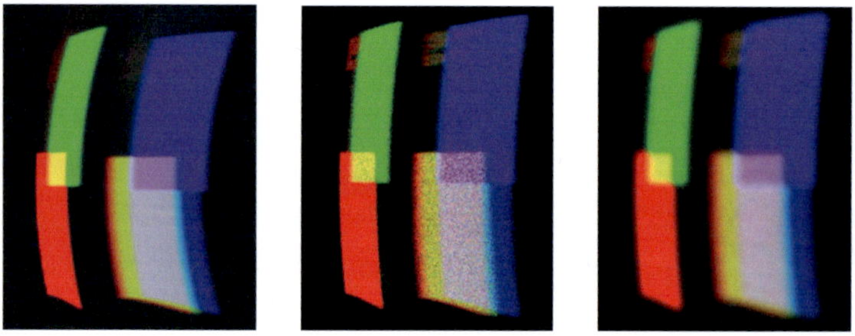

(a) (b) (c)

Figure 8.25: Comparison of far field measurement (a) and unfiltered (b) as well as filtered (c) spectral ray tracing with Red/Green/Blue/White ray file.

The main part of the wrongly assigned flux from the green LED of Fig. 8.20 is assigned to the white LED and does not disturb the obtained images in Fig. 8.25 (b) and (c). Thus, the comparison of Fig. 8.25 (a) and Fig. 8.25 (c) shows a good qualitative agreement between the simulation results of the RGBW source with the prism in its near field and the measured data. Note that the slight red and white chromatic shadows at the green and blue distributions are also caused by a reflection of the specific radiation at the prism.

However, the curvature of the distributions seems again more pronounced for the measurement. Also note that in the measurement the green distribution is located slightly above the blue distribution, which is not true for the simulation. This again indicates a misalignment between the far field measurement and the simulation or the simulation and the near field measurement. Furthermore, a small green color shadow can be noted at the lower right part of the red LED in the non-filtered simulation in Fig. 8.25 (b). This is a reconstruction artifact of PMBS, which occurred due to the wrongly assigned flux in the green LED model. Appendix B shows that - in contrast to the green shadow - the slight misalignment between measurement and simulation is not caused by the spectral ray files created with PMBS.

A quantitative comparison in the form of the chromaticity distance is shown in Fig. 8.26. Figure 8.26 (a) visualizes the chromaticity distance for each measured location, which had a suitable measurement signal in at least two of the three measurement channels of the colorimeter. The same applies for the simulation. Furthermore, the distribution of all deviations is shown as a histogram in Fig. 8.26 (b). The figure also summarizes the mean chromaticity distance of the six highlighted regions shown in Fig. 8.26 (a).

The highest deviations are again located at the edges of the distribution. High deviations also occur at the edges of the overlapping regions Y and C but not within the overlapping regions Y and C. This also indicates the slight misalignment between the measured and the simulated distributions. Consequentially, the highest deviations are located at the lower right corner as well as the upper right corner of the white LED region, which has the strongest difference in the curvature.

The XMP-filter and the number of rays used in combination with the chromatic shift of the prism may also be critical for the white region as described in the previous example. Furthermore, only 15% of the blue LED model is assigned to the blue chip of the white LED. This also decreases the number of rays with short wavelengths in the white region. Note that these deviations are by far the highest.

Nevertheless, as the overall mean chromaticity distance is influenced by these deviations at the edges, it is not suited to validate the spectral ray files. A better quantitative validation is provided by the mean values of the six locations G, Y, R, B, C and W, which are less influenced by the misalignment. In that case, the highest mean chromaticity distance of a single spectral source occurs in the red region R and the white region W. However, both can be expected. As the basis spectra model used was a Logistic power peak, the deviations between the red LED basis spectrum and the true red LED spectrum are in this order of magnitude (see Table 5.2 in chapter 5). The deviations in the white region may occur due to the non-optimal basis spectra of the phosphor, the wrongly added flux parts of the green LED model and due to noise, which is stronger in this region compared to other regions. The main reasons for the deviations in the region Y is the deviation of the red region and the 5% missing flux in the green LED model, which disturbs the relative weighting of the color mixture.

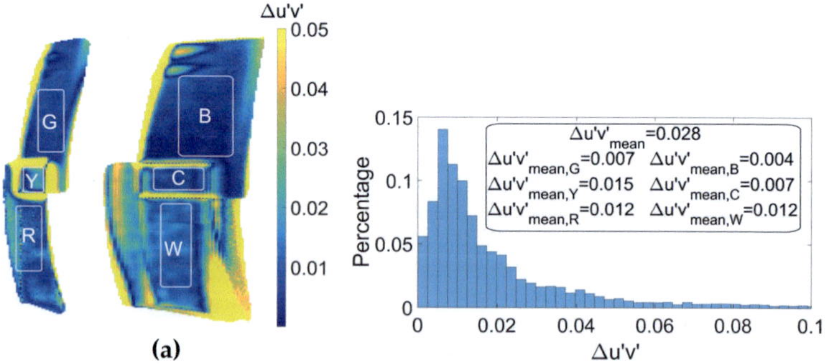

(a) Pixel-wise **(b)** Distribution

Figure 8.26: Quantitative comparison of measurement and spectral ray tracing with Red/Green/Blue/White ray file.

Even so, the absolute deviations are again within the color discrimination thresholds in practical applications according to [62]. However, it should be noted that these thresholds depend on the specific viewing conditions and were determined for white light colors. Recent studies [107], [108] hint that the color discrimination threshold is higher and thus less critical for saturated colors. As the regions R, W and Y all have an offset induced error, the relative deviations between measurement and simulation, which are required for a uniformity analysis, are even smaller.

In summary, both tested spectral ray files - Red/White and RGBW - are well suited for optical simulations. Apart from the green shadow in the RGBW distribution and slight colorimetric offsets below $\Delta u'v' < 0.01$, no PMBS induced artifacts were observed. As these spectral reconstructions are more complicated than a single phosphor converted white LED or a RGB LED, it can be concluded that PMBS can be used directly for all these LED systems to simulate spectral distributions in commercial ray tracing software with an appropriate consideration of the spectral near field of these LED systems.

CHAPTER 9

DISCUSSION

In this thesis, a measurement and modeling technique to obtain hyperspectral LED models is proposed, implemented and verified. The first section of this chapter provides a theoretical discussion of the overall approach. The second section focuses on practical aspects. This includes an analysis of the achieved results and the required measurement time - also compared to the state of the art - and concludes with practical applications.

9.1 PMBS – APPROACH

The physically motivated basis spectra approach (PMBS) is a method to create hyperspectral LED models based on the assumption that all spectral variations in the near field of a LED source can be described as the weighted sum of their physical basis spectra. The relative spectral distribution of each spectral source is constant. Therefore, PMBS utilizes prior knowledge regarding the device under test.

In the best case, this prior knowledge includes the spectral distributions of the physical sources of the device under test. In the worst case, the basis spectra have to be estimated. Then the only important, but easily obtainable information is the number of different spectral sources in the device. Furthermore, PMBS requires detailed knowledge regarding the spectral sensitivity of the ILMD-based measurement system. Subsequently, all available spectral information is combined to set up a system of linear equations, which is used to reconstruct the individual spectral models from the ILMD-based measurements. The combination of all spectral models is the hyperspectral model.

In a similar way to the general approach, the proposed practical implementations aim to utilize physical boundary conditions. Examples are the assumption that colored LEDs are spatially separated from each other or that negative amplitudes are physically impossible. The validity of these assumptions is unquestionable. Nevertheless, there are theoretical limitations in the overall approach or specific aspects which have to be discussed in order to assess the developed approach correctly.

The first aspect to be considered is the Monte Carlo simulation used for the filter selection, which aims to minimize the expected reconstruction errors described in section 6.2. By modeling typical uncertainties of the ILMD-based measurements in relation to the optical filters and spectral sensitivity of the ILMD, their influence on the spectral reconstruction is estimated. This information is used to select an optimal filter set. This modeling requires basis spectra, which are at this point not yet optimized. However, as indicated in section 6.4, the filter selection using standard filters and glass edge absorption filters - generally filters with a broad transmission - is stable towards non-optimized basis spectra. The Monte Carlo simulation can therefore be used as it is for choosing an optimal or at least one of the best filter combinations. The determination of the best filter combination cannot be guaranteed, because the number of roughly estimated input parameters influences the result. But performing the filter selection will always minimize the risk of selecting an unsuitable filter combination in the context of PMBS.

Nevertheless, it has to be noted that although the Monte Carlo simulation selects the filters based on estimated reconstruction errors regarding measurement uncertainties, an estimation of the absolute reconstruction uncertainties is far more complex. This would require an analysis according to [93], which requires modeling the input uncertainties and also the device under test in more detail. An example is the prior estimation of the amplitude distributions of all spectral sources or influences of environmental conditions such as temperature or current on the specific device under test. However, predicting absolute uncertainties is, in contrast to their general consideration, not within the scope of this application.

The second aspect is the calculation procedure of chapter 7 used to apply PMBS directly to commercially available ray files. In the context of PMBS, the histogram based models and the calculation workflow are only a means to an end. They might be adapted, and other quantification models, raw data such as .ttr-files, or ILMD-based images can be used as well. However, it is important to note that the usability of those algorithms is higher, if only a small amount and only intuitive input parameters are used. In this case these were the desired NCC and the final amount of negative amplitudes $Q_{\text{final},n}$.

The core concept of the calculation procedure with respect to PMBS is the nearest neighbor calculation of subsection 7.3.2. It utilizes the information that negative spectral distributions of physical sources are impossible. In theory, the nearest neighbor calculation allows maintaining the resolution of the input data by assuming that all negative amplitudes occur due to a resolution caused mismatch.

However, both negative and false positive amplitudes can also be induced either by deviating transmission filter profiles or by an insufficient description of the basis spectra. The occurrence of these initial reconstruction errors in the form of negative amplitudes has to be checked prior to the calculation procedure such that the nearest neighbor calculation only converges towards the initial amount of negative amplitudes. Otherwise, the assumption that only resolution caused negative values are canceled out is no longer valid. If initial spectral reconstruction errors are minimized with the nearest neighbor calculation, the calculation procedure may require an unnecessary additional amount of calculation time and create additional artifacts. The initial amount of negative amplitudes was considered during the reconstruction of the RGBW example in section 8.2.

Furthermore, it is important to note that it cannot be guaranteed that only false positive amplitudes are used to compensate the negative amplitudes. Further, the nearest neighbor calculation can only be used with physically motivated basis spectra because the usage of negative amplitudes requires the prior knowledge that they have to be calculation artifacts. Mathematical basis spectra, which would for instance occur if a principal component analysis were used to derive the basis spectra, may also create correct negative amplitudes within an individual model. They only compensate each other after the final combination of all models.

The third aspect is the basis spectra optimization using spatial separation. The spatial separation utilizes the information that it is very unlikely for the same spatial location to emit, for instance, a red, green and blue LED spectrum at once. It is assumed that colored LED are spatially separated from each other. As outlined in subsection 5.2.3 this technique is only necessary if the basis spectra are not available and are hard to estimate due to a significant spectral overlap. An example is the combination of green LEDs and a yellow phosphor as in subsection 5.2.3 and section 8.2. The basis spectra optimization uses the spectral reconstruction of PMBS to create irradiance images and assesses the results regarding their spatial separation. The optimization changes the basis spectra to achieve a better spatial separation. However, this will only work up to a certain degree.

The basis spectra optimization only assesses the spatial separation based on a PMBS reconstruction and therefore only validates the reconstruction matrix. The basis spectra are only validated indirectly. This has three consequences regarding PMBS.

The first is that the relative weightings have to be based on values, which are independent on the basis spectra. An example is the camera luminous flux provided by the RIGO801. The more elegant inverse reconstruction cannot be used because it assumes correct basis spectra.

The second consequence is that deviations from at least one measured spectrum have to be included as an auxiliary condition while effectively changing the reconstruction matrix. This was also considered in all examples which included the basis spectra optimization. In complex cases adding a low number of additional spectral measurements is recommended as the measurement time is not enlarged significantly while the auxiliary condition becomes more powerful. However, it must be ensured that all angular positions used in the auxiliary conditions have a suitable signal to noise ratio. This applies for the ILMD-based measurements and for the spectral measurement.

The third consequence affects the filter selection. Filters with a broad transmission are more suited to use the spatial separation as figure of merit for the basis spectra optimization. The worst case for the spatial separation would be a reconstruction matrix which equals the unity matrix because then the spatial separation is already optimal and the non-optimized basis spectra have no effect on the spatial separation.

The fourth aspect is the most general. All modeling concepts require the tacit assumption that the complete system has no additional inaccuracies besides basis spectra deviations or resolution caused artifacts. These two problems are directly related to PMBS and can be considered the main sources of spectral reconstruction errors. Therefore, they are corrected and optimized during the PMBS workflow. However, there are additional aspects influencing the reconstruction results. These are not solely related to PMBS but to near field photometry in general. The optical filter and neutral density filter transmission profiles are first examples. Although they are considered during the filter selection, their influence is only minimized but not corrected. Other inaccuracies are, for instance, the lens calibration and aberrations, or the spectral measurement as well as an unstable device under test.

With this in mind, the last figure of this thesis can be interpreted. Figure 9.1 shows the spatial separation of the RGBW reconstruction of section 8.2 using the individually measured basis spectra and an optimized filter set at a low resolution. This means that all aspects of this thesis are optimized. However, there are still artifacts within the green LED model, which are caused by other factors. The artifacts are even slightly larger than the artifacts in the spatial separation of the optimized model of Fig. 8.19.

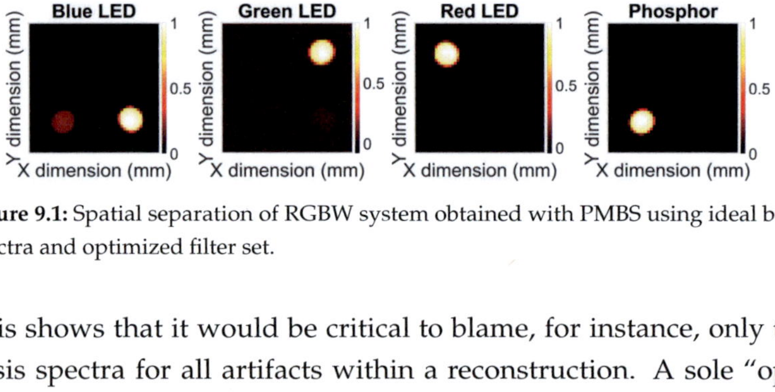

Figure 9.1: Spatial separation of RGBW system obtained with PMBS using ideal basis spectra and optimized filter set.

This shows that it would be critical to blame, for instance, only the basis spectra for all artifacts within a reconstruction. A sole "optimization" of the basis spectra would only transfer all ignored aspects directly into the basis spectra. Generally, a sole optimization of the main challenges of PMBS, the basis spectra estimation, the filter selection, and the calculation procedure cannot lead to hyperspectral LED models without any artifacts because they are also caused by factors independent of PMBS. However, hyperspectral LED modeling using PMBS will improve together with near field photometry and may even help to improve this aspect. This will be considered in more detail in the outlook of this thesis.

9.2 PMBS – PRACTICAL APPLICATIONS

The PMBS approach has been developed to be a fast modeling technique for spectral near field data of LEDs with a low spectral error. The measurement and simulation effort has to remain feasible with respect to the current state of the art. Furthermore, the complexity of spectral reconstructions depends on the specific LED system.

One of simplest, but most important cases is the phosphor converted white LED. Most state-of-the-art techniques focus mainly on this case. The industrial applied quasi-standard is the Blue/Yellow approach, which combines a short measurement time with reasonable results regarding the spectral error. The basis spectra importance is low compared to spatially separated different colored LEDs because all angular and spatial regions provide a mixture of the blue LED spectrum and the yellow phosphor spectrum.

PMBS, which can be interpreted as the physical generalization of the Blue/Yellow approach, requires the same measurement effort for phosphor converted white LEDs. It consists of only two ILMD-based measurements and one spectral measurement. The additional simulation time of PMBS is smaller than the measurement time. Section 8.1 shows a direct comparison of both approaches. Both the relative color over angle deviations as well as the absolute chromaticity distances are smaller and even in the region of just noticeable color differences for PMBS. Also, only standard filters are required to successfully apply

PMBS. Therefore, PMBS is well suited to create spectral ray files for the important case of phosphor converted white LEDs.

Note that neither non-linear effects such as quenching within the phosphor conversion nor inhomogeneous combinations of different phosphors are covered in PMBS or the Blue/Yellow approach. However, as those effects have been, and will be, considered in the development of phosphor converted LEDs [75], [109]–[111], their practical relevance is limited.

Another simple case is the RGB LED, which was reconstructed in section 7.5 to verify the calculation procedure using only three ILMD-based measurements with standard filters. The basis spectra modeling without a phosphor is uncritical. Depending on the desired precision of the calculation procedure, the reconstructed spectral models provide details such as the electrode structure of the LEDs.

A more complicated, but practical case is the combination of red LEDs and phosphor converted white LEDs. The polychromatic ray data is the only state-of-the-art method, which covered this case by combining eight ILMD-based measurements with angularly resolved spectral measurements [61]. The PMBS method covered this case in section 8.2 by applying only three ILMD-based measurements and one spectral measurement significantly reducing the measurement time. As in the case of the phosphor converted white and RGB LED, standard filters remain suitable for applying PMBS. The initial modeling provides well suited estimates for the basis spectra. This was also the first case validated by spectral ray tracing. The highest mean chromaticity distance between the simulations and the colorimeter measurements was only $\Delta u'v' \approx 0.013$. A large part of the deviations was an offset induced error of $\Delta u'v' \approx 0.009$. Therefore, the reconstructed spectral ray files are well suited for spectral ray tracing and uniformity analyses.

The most complex case considered in this thesis is the RGBW example in section 8.2. Although some state-of-the-art methods can be used in theory, a spectral ray file of such a system had not been previously documented. It is the only example in this thesis which requires a basis spectra optimization. In addition, the requirements for the quality of the input data, for example the transmission filter profiles, increase. Therefore, it is necessary to select an optimal filter combination in order to minimize the influence of uncertainties outside the scope of PMBS. Despite all optimizations, the spectral ray tracing still showed slight PMBS induced artifacts such as a tiny green color shadow within the red part of the distribution. Nevertheless, due to the advanced modeling techniques of PMBS such as the basis spectra optimization, the quantitative results are comparable to those of Red/White and thus well suited for spectral ray tracing and uniformity analyses. Furthermore, the quantitative analysis of the spectral ray tracing considered neither the uncertainties of the array spectroradiometer and the colorimeter, nor the fact that the CIE $u'v'$ diagram is more reliable for white color stimuli [42], [43]. Physiological studies suggest that the noticeable color differences in the CIE $u'v'$ diagram increase for saturated colors [107], [108], which would decrease the importance of the offset induced deviations in these regions.

The only practical limitation of the PMBS approach as described in this thesis is the combination of different white LEDs, which use different phosphors. An example is a warm white/cold white LED system, which uses a different phosphor for the warm white and cold white LEDs. The main limitation is the initial modeling of the basis spectra. The deduced figures of merit can be used but the basis spectra optimization requires an adaption of the optimization algorithm with respect to the phosphor distributions. However, if the basis spectra are known, a hyperspectral LED model can be created with PMBS because both the filter selection as well as the calculation procedure have no fundamental limitations.

As described above and in subsection 3.4.7, PMBS has significant advantages compared to other state-of-the-art methods. It is, for example, suitable for several different LED systems which was validated in this thesis. Other state-of-the-art methods are only suitable for phosphor converted white LEDs or less complicated systems. Methods which are able to create spectral ray files of more complicated LED systems require angularly resolved spectral measurements. This is time-consuming especially compared to the short ILMD-based measurement time of PMBS.

However, PMBS is only suited for LED systems. In the case of light sources without a spatially-spectral variation such as OLEDs the PCA approach provides good results. In the case of very complex systems, as for example a combination of different OLEDs and different LEDs, a time-consuming approach such as precise physical modeling or the polychromatic ray data approach, which requires no specific modeling assumptions, should be used.

To sum up, the developed approach was successfully used to obtain hyperspectral models of phosphor converted white LED, RGB LED and Red/White LEDs with a comparably small measurement and simulation effort using only standard optical filters. In these cases no additional measurement equipment is required. More complex cases require an optimized filter set in order to minimize the influence of measurement uncertainties and the application of advanced modeling techniques such as the proposed basis spectra optimization.

CHAPTER 10

SUMMARY

This chapter summarizes the developed method and achieved results and provides an outlook regarding potential improvements. Furthermore, it suggests applications outside the original scope of the thesis.

10.1 SUMMARY

The aim of the thesis was to propose, develop and validate a measurement and modeling technique for obtaining the spectral near field of LEDs in order to improve the optical design of high quality LED based lighting systems. Based on selected concepts of the current state of the art, the physically motivated basis spectra (PMBS) approach, which considers the required measurement effort as well as the usability and generality in practical applications, was proposed and mathematically described.

The main assumption of PMBS is that spectral variations in the near field of an LED system can be described as the weighted sum of all physically occurring spectral sources. This enables the possibility of interpreting a hyperspectral LED model as a set of spectral models, which can each be described individually, for instance as a ray file with one global spectral distribution. The main measurement effort in the form of conventional ILMD-based (imaging luminance measurement device) goniophotometric measurements with different optical filters is determined by the number of spectral sources and is therefore, in contrast to the state of the art, always minimized. The ILMD-based measurements and one global spectral measurement as well as a detailed spectral description of the ILMD-based measurement system are used to reconstruct the near field distribution of the individual spectral models.

The thesis then defined three main challenges concerning the practical implementation of PMBS. These were the potentially mandatory basis spectra estimation, reconstruction errors caused by deviating input data and the application of PMBS on typically available near field data.

The basis spectra estimation uses the global spectral measurement and the number of different spectral sources in the system to obtain phenomenological LED basis spectra. If necessary, the basis spectra can be optimized by quantifying the "spatial separation". This figure of merit assumes that single colored chips are both spatially separated and highly localized in the three dimensional space. In other words, it assumes that a red LED is only red, a green LED is only green, a blue LED is only blue and a white LED is only white or rather blue and yellow. In the case of the spatially non-separated individual phosphor converted white LEDs, a color mixing line, which compares possible color mixtures to actual measurements, can be used. These techniques enable the estimation of the required basis spectra, if they are not available in advance.

Furthermore, it has been shown that an unintelligent filter selection can lead to undesired reconstruction artifacts due to given uncertainties regarding the ILMD-system. To correctly assess the influence of the selected optical filters on PMBS, empirical models for common glass edge absorption and interference filters as provided by the manufacturers were extended to consider physically caused measurement uncertainties. While these models may be evaluated directly in a Monte Carlo simulation, a time saving preselection technique using factorial design was also proposed and validated. During the validation of the filter selection techniques, it was verified that an appropriate filter set is able to minimize reconstruction artifacts. Furthermore, it was shown that the procedure is stable with respect to estimated or rather non-optimized basis spectra, especially for optical filters with a broad transmission. Practical examples are glass edge absorption filters.

To ensure the direct usability of PMBS, the proposed and validated calculation procedure focused on commercially available ray files. The ray files are transformed into histogram based light source models, whose resolution parameters are adapted with respect to a user defined precision parameter. Calculation artifacts, which are caused by the highly discrete nature of ray files, are corrected by applying the nearest neighbor calculation, which utilizes the physical impossibility of physically motivated spectral sources with negative or partly negative amplitudes. This process ensures a correct application of PMBS with high model resolutions. Furthermore, the obtained results can be used directly in state-of-the-art ray tracers.

Using the described implementation, PMBS was first used to reconstruct a conventional phosphor converted white LED. The required measurement effort is the same as in the Blue/Yellow approach, which considers only this type of LED. The hyperspectral LED model, which was obtained with PMBS, showed smaller deviations from validation measurements. The obtained chromaticity distances were even in the region of just noticeable color differences at around $\Delta u'v' \approx 0.002$.

PMBS was further tested by successfully modeling a RGB LED, which covered details such as the electrode structure, as well as spectral ray files of more complex Red/White and Red/Green/Blue/White LED-systems. The spectral ray files were used within a commercial ray tracer to obtain a spectral far field distribution, created by a spectrally sensitive prism in the near field of the measurement object. The spectral simulations were compared to measured colorimetric values of the same setup and showed deviations below $\Delta u'v' \leq 0.016$, which led to similar images of these distributions. Therefore, it can be concluded that PMBS is suitable for obtaining hyperspectral LED models, which can be used to improve the optical design of high quality LED based lighting systems.

10.2 OUTLOOK

Although PMBS obtains good results for nearly all LED systems, there is still a practical limitation. The application of PMBS on LED systems with different phosphors on different LEDs is only possible, if the basis spectra are known in advance because the initial basis spectra estimation is unsuited for this case. The proposed figure of merit spatial separation remains valid and may still be used in optimizations. However, these optimizations require the development of an appropriate spectral adaptation algorithm, which may for instance try to preserve specific spectral features of the individual phosphor spectra, similar to the spectral adjusting method. Another option to obtain first estimates of the basis spectra is to base the complete initial basis spectra modeling on a single or on a small number of hyperspectral measurements.

The conventional ray files used as input data in this thesis, are already processed data and therefore limit the resolution from the start. Furthermore, the adaptation of the resolution parameters with respect to the input ray files is the bottle neck regarding the computation time. A higher resolution as well as a faster computation time might be achieved by applying PMBS directly on raw data as for instance compressed luminance images as .ttr-files. While the data compression and adaptive resolution estimation is avoided, the nearest neighbor calculations will still remain the core concept of each calculation procedure.

However, PMBS is not limited to obtaining spectral near field data for optical simulations. There might be other applications as well. In particular, the fact that the reconstruction of phosphor converted white LEDs, RGB LEDs and Red/White LEDs requires only standard filters is a promising result. It enables the possibility of estimating a spectral light intensity distribution curve of these LED systems based solely on colorimetric measurements. Furthermore, it is possible to use PMBS to reconstruct a spectral image of LED systems and LED based illuminations with non-goniometric ILMD-based images. If the measurement time is small, it might also be advantageous to consider an over-determined system of linear equations in PMBS.

In this thesis, it was always assumed that all spectral transmissions and sensitivities in the measurement system as well as all measured input data used in the reconstruction were correct. Only the spectral near field of the device under test was unknown. However, this assumption can be inverted by using a spatially, angularly and spectrally well described LED source as developed in the metrology project EMRP ENG62 MESaIL. Then the reconstructions of PMBS, which relate spectral measurements to several ILMD-based near field measurements, might be used to qualify near field measurements and thus near field measurement systems. This may help to advance near field measurements beyond spectral simulations but in the context of photometry in the future.

BIBLIOGRAPHY

[1] N. Holonyak Jr and S. Bevacqua, "Coherent (visible) light emission from Ga (As1- xPx) junctions", *Applied Physics Letters*, vol. 1, no. 4, pp. 82–83, 1962.

[2] E. F. Schubert and J. K. Kim, "Solid-state light sources getting smart", *Science*, vol. 308, no. 5726, pp. 1274–1278, 2005.

[3] T. M. Katona, P. M. Pattison, and S. Paolini, "Status of Solid State Lighting Product Development and Future Trends for General Illumination", *Annual Review of Chemical and Biomolecular Engineering*, vol. 7, no. 1, pp. 263–281, 2016.

[4] S. Nakamura, T. Mukai, and M. Senoh, "Candela-class high-brightness InGaN/AlGaN double-heterostructure blue-light-emitting diodes", *Applied Physics Letters*, vol. 64, no. 13, pp. 1687–1689, 1994.

[5] T. Taguchi, Y. Uchida, and K. Kobashi, "Efficient white LED lighting and its application to medical fields", *physica status solidi (a)*, vol. 201, no. 12, pp. 2730–2735, 2004.

[6] K. Kobashi and T. Taguchi, "Warm white LEDs lighting over Ra=95 and its applications", in *Proc. SPIE*, vol. 6486, 2007.

[7] R. C. Morrow, "LED Lighting in Horticulture", *HortScience*, vol. 43, no. 7, pp. 1947–1950, 2008.

[8] J. Bhardwaj, R. Peddada, and B. Spinger, "Advances in LEDs for automotive applications", in *Proc. SPIE*, vol. 9768, 2016.

[9] A. D. Almeida, B. Santos, B. Paolo, and M. Quicheron, "Solid state lighting review - Potential and challenges in Europe", *Re-*

newable and Sustainable Energy Reviews, vol. 34, no. Supplement C, pp. 30 –48, 2014.

[10] A. Nardelli, E. Deuschle, L. D. de Azevedo, J. L. N. Pessoa, and E. Ghisi, "Assessment of Light Emitting Diodes technology for general lighting: A critical review", *Renewable and Sustainable Energy Reviews*, vol. 75, no. Supplement C, pp. 368–379, 2017.

[11] T. Kuntze, "Plastic Optics Enable LED Lighting Revolution", *Optik & Photonik*, vol. 2, no. 4, pp. 42–45, 2007.

[12] S. Wendel, "Freiform-Optiken im Nahfeld von LEDs", Spektrum der Lichttechnik Band 7. KIT Scientific Publishing, 2014, PhD thesis, Karlsruhe Institute of Technology.

[13] K. Wang, F. Chen, Z. Liu, X. Luo, and S. Liu, "Design of compact freeform lens for application specific light-emitting diode packaging", *Optics Express*, vol. 18, no. 2, pp. 413–425, 2010.

[14] I. Ashdown, "Near-Field Photometry: A New Approach", *Journal of the Illuminating Engineering Society*, vol. 22, no. 1, pp. 163–180, 1993.

[15] I. Ashdown and M. Salsbury, "A near-field goniospectroradiometer for LED measurements", in *Proc. SPIE*, vol. 6342, 2006.

[16] R. Dürr and U. Streppel, "Rayfiles for non-sequential raytracing", in *Proceedings of the International Light Simulation Symposium*, 2012, pp. 11–24.

[17] Illuminating Engineering Society, *IES TM25-13 Ray File Format for the Description of the Emission Property of Light Sources*, 2013.

[18] H.-J. J. Frasch and G. Steinmetz, "Monte Carlo methods in raytracing software", in *Proc. SPIE*, vol. 4769, 2002, pp. 54–66.

[19] P. R. Boyce, *Human Factors in Lighting*. CRC Press, 2014.

[20] K. Bieske and O. Dierbach, "Evaluation des Einsatzes von tageslichtähnlichem Kunstlicht in der gerontopsychiatrischen Pflege

und Betreuung Hochbetagter", in *5. Symposium Licht und Gesundheit*, 2006.

[21] C. A. Sust, P. Dehoff, C. Hallwirth-Spörk, D. Lang, and D. Lorenz, "Mehr Licht! Verbesserung des Wohlbefindens durch biologisch wirksames Licht bei Demenzkranken", in *Licht*, 2014.

[22] K. Choi and H.-J. Suk, "Dynamic lighting system for the learning environment: Performance of elementary students", *Optics Express*, vol. 24, no. 10, A907–A916, 2016.

[23] Y. Ohno, "Spectral design considerations for white LED color rendering", *Optical Engineering*, vol. 44, no. 11, pp. 111 302–1–111302–9, 2005.

[24] P. Zhong, G. He, and M. Zhang, "Spectral optimization of the color temperature tunable white light-emitting diode (LED) cluster consisting of direct-emission blue and red LEDs and a diphosphor conversion LED", *Optics Express*, vol. 20, no. 105, A684–A693, 2012.

[25] F. Zhang, H. Xu, and Z. Wang, "Spectral design methods for multi-channel LED light sources based on differential evolution", *Applied Optics*, vol. 55, no. 28, pp. 7771–7781, 2016.

[26] F. Zhang, H. Xu, and Z. Wang, "Optimizing spectral compositions of multichannel LED light sources by IES color fidelity index and luminous efficacy of radiation", *Applied Optics*, vol. 56, no. 7, pp. 1962–1971, 2017.

[27] T. S. Bonenberger, J. Baumgart, and C. Neumann, "Angular and spatial color mixing using mixing rods with the geometry of a chaotic-dispersive billiard system", *Advanced Optical Technologies*, vol. 5, no. 2, pp. 157–165, 2016.

[28] M. Chakrabarti, H. C. Pedersen, P. M. Petersen, C. Poulsen, P. B. Poulsen, and C. Dam-Hansen, "High-flux focusable color-tunable and efficient white-light-emitting diode light engine for

stage lighting", *Optical Engineering*, vol. 55, no. 8, pp. 085 101–1–085101–12, 2016.

[29] H.-C. Wang, Y.-T. Chiang, C.-Y. Lin, M.-Y. Lu, M. K. Lee, S.-W. Feng, and C.-T. Kuo, "All-reflective RGB LED flashlight design for effective color mixing", *Optics Express*, vol. 24, no. 5, pp. 4411–4420, 2016.

[30] G. Durinck, F. B. Leloup, J. Audenaert, and P. Hanselaer, "Modeling the spectrum of a luminaire including a dichroic filter by spectral ray tracing", *Light & Engineering*, vol. 25, no. 2, 2017.

[31] R. Windisch, G. Heidel, U. Binder, and K. Bergenek, "Impact of spectral features of common LED lighting systems on TM-30 color indices", *Optics Express*, vol. 25, no. 3, pp. 1824–1830, 2017.

[32] OSRAM Opto Semiconductors GmbH, *OSLON SSL 150 Datasheet GF CSHPM1.24*, Version 1.1, 2015.

[33] OSRAM Opto Semiconductors GmbH, *TOPLED Black Datasheet LO T64F*, Version 2.4 (OS-PCN-2016-025-A), 2016.

[34] OSRAM Opto Semiconductors GmbH, *OSRAM OSTAR Stage Datasheet LE RTDUW S2WN*, Version 1.0, 2017.

[35] OSRAM Opto Semiconductors GmbH, *OSLON Signal Datasheet LV CQBP*, Version 1.1, 2015.

[36] S. Liu and X. Luo, *LED Packaging for Lighting Applications: Design, Manufacturing and Testing*. John Wiley & Sons (Asia), 2011.

[37] T. Q. Khanh, P. Bodrogi, T. Vinh Quang, and H. Winkler, *LED Lighting : Technology and Perception*. Wiley-VCH, 2015.

[38] OSRAM Opto Semiconductors GmbH, *Golden DRAGON Datasheet LCW W5SM*, Version 1.3, 2017.

[39] OSRAM Opto Semiconductors GmbH, *Golden DRAGON Datasheet LUW W5SM*, Version 1.1, 2017.

[40] J. M. Palmer and B. G. Grant, *The Art of Radiometry*. SPIE Press, 2010.

[41] W. R. McCluney, *Introduction to Radiometry and Photometry*, 2. ed. Artech House, 2014.

[42] M. Luo, G. Cui, and M. Georgoula, "Colour difference evaluation for white light sources", *Lighting Research & Technology*, vol. 47, no. 3, pp. 360–369, 2015.

[43] Commission internationale de l'Eclairage, *CIE TN 001:2014: Chromaticity Difference Specification for Light Sources*, 2014.

[44] T. Q. Khanh, P. Bodrogi, and T. Q. Vinh, *Color Quality of Semiconductor and Conventional Light Sources*. John Wiley & Sons, 2017.

[45] G. Wyszecki and W. S. Stiles, *Color science: Concepts and Methods, Quantitative Data and Formulae*, 2. ed. Wiley, 2000.

[46] M. D. Fairchild, *Color Appearance Models*, 2. ed. Wiley, 2006.

[47] B. E. A. Saleh and M. C. Teich, *Grundlagen der Photonik*, 2. ed. John Wiley & Sons, 2008.

[48] D. Malacara and Z. Malacara, *Handbook of Optical Design*, 2. ed. Dekker, 2004.

[49] T. Werner, "Simulation, Aufbau und Charakterisierung von autostereoskopischen Display-Systemen im Fahrzeugbereich", Spektrum der Lichttechnik Band 13. KIT Scientific Publishing, 2017, PhD thesis, Karlsruhe Institute of Technology.

[50] V. Jacobs, P. Blattner, Y. Ohno, U. Krüger, P. Hanselaer, P. Rombauts, and F. Schmidt, "Analysis of errors associated with photometric distance in goniophotometry", in *Proceedings of the 28th Session of the CIE*, 2015.

[51] A. Domhardt, "Analytisches design von freiformoptiken für punktlichtquellen", Spektrum der Lichttechnik Band 5. KIT Scientific Publishing, 2013, PhD thesis, Karlsruhe Institute of Technology, Karlsruhe.

[52] I. L. Sayanca, K. Trampert, and C. Neumann, "Indirect Light Intensity Distribution Measurement using Image Merging", in

Advanced Mathematical and Computational Tools in Metrology and Testing, 2017.

[53] J. Audenaert, G. Durinck, F. B. Leloup, G. Deconinck, and P. Hanselaer, "Simulating the spatial luminance distribution of planar light sources by sampling of ray files", *Optics Express*, vol. 21, no. 20, pp. 24 099–24 111, 2013.

[54] S. Häring, "Erweiterung des simulationsprozesses von lichtfeldern für die virtuelle leuchtenentwicklung im automobilbau", PhD thesis, Ilmenau University of Technology, 2009.

[55] G. G. Gregory, I. Ashdown, W. Brandenburg, D. Chabaud, O. Dross, S. Gangadhara, K. Garcia, M. Gauvin, D. Hansen, K. Haraguchi, G. Hasna, J. Jiao, R. Kelley, J. Koshel, and J. Muschaweck, "Data format standard for sharing light source measurements", in *Proc. SPIE*, vol. 8835, 2013.

[56] K. Bredemeier, F. Schmidt, and W. Jordanov, "Ray Data of LEDs and Arc Lamps", in *International Symposium on Automotive Lighting*, 2005.

[57] TechnoTeam Bildverarbeitung GmbH, *Prospekt RIGO801 Near-Field Goniophotometer*, Version 2.0, 2017.

[58] TechnoTeam Bildverarbeitung GmbH, *LMK 50 Color datasheet*, 2012, (accessed July 2015).

[59] G. Leschhorn and R. Young, *Handbook of LED and SSL Metrology*. Instrument Systems GmbH, 2017.

[60] E. H. Adelson and J. R. Bergen, "The Plenoptic Function and the Elements of Early Vision", in *Computational Models of Visual Processing*, 1991, pp. 3–20.

[61] D. Hansen and V. Schumacher, "Spectral Raydata for Simulation of Color Rendering Indices", in *LED Professional Symposium*, 2013.

[62] K. Bieske and T. Fiebig, "Kombination unterschiedlicher Lichtfarben im Raum: Farbige Schatten", in *Licht*, 2014.

[63] R. Rykowsky, "Spectral ray tracing from near field goniophotometer measurements", *Light & Engineering*, vol. 19, no. 1, pp. 23–29, 2011.

[64] D. Hansen, "Messung und Simulation polychromatischer Strahldaten", *Automobiltechnische Zeitschrift*, vol. 114, no. 11, pp. 869–873, 2012.

[65] D. Hansen, S. Paul, and V. Schumacher, "LED Measurement to Obtain Polychromatic Raydata and their Value for Simulations", in *LED Professional Symposium*, 2012.

[66] V. A. Jacobs, J. Audenaert, J. Bleumers, G. Durinck, P. Rombauts, and P. Hanselaer, "Rayfiles including spectral and colorimetric information", *Optics Express*, vol. 23, no. 7, A361–A370, 2015.

[67] OSRAM Opto Semiconductors GmbH, *Application Note: Importing Rayfiles of LED from OSRAM Opto Semiconductors*, 2013.

[68] T.-X. Lee, T.-L. Lu, and B.-S. Chen, "Combining near-field hyperspectral imaging and far-field spectral-angular distribution to develop mid-field white LED optical models with spatial color deviation", *Optics Express*, vol. 24, no. 14, A1094–A1106, 2016.

[69] C.-C. Sun, T.-X. Lee, S.-H. Ma, Y.-L. Lee, and S.-M. Huang, "Precise optical modeling for LED lighting verified by cross correlation in the midfield region", *Optics Letters*, vol. 31, no. 14, pp. 2193–2195, 2006.

[70] C. Sommer, P. Hartmann, P. Pachler, M. Schweighart, S. Tasch, G. Leising, and F. P. Wenzl, "A detailed study on the requirements for angular homogeneity of phosphor converted high power white LED light sources", *Optical Materials*, vol. 31, no. 6, pp. 837–848, 2009.

[71] S.-L. Hsiao, N.-C. Hu, and H. Cornelissen, "Phosphor-converted LED modeling using near-field chromatic luminance data", *Optics Express*, vol. 21, no. S2, A250–A261, 2013.

[72] C.-C. Sun, C.-Y. Chen, C.-C. Chen, C.-Y. Chiu, Y.-N. Peng, Y.-H. Wang, T.-H. Yang, T.-Y. Chung, and C.-Y. Chung, "High uniformity in angular correlated-color-temperature distribution of white LEDs from 2800K to 6500K", *Optics Express*, vol. 20, no. 6, pp. 6622–6630, 2012.

[73] C.-C. Sun, C.-Y. Chen, H.-Y. He, C.-C. Chen, W.-T. Chien, T.-X. Lee, and T.-H. Yang, "Precise optical modeling for silicate-based white LEDs", *Optics Express*, vol. 16, no. 24, pp. 20 060–20 066, 2008.

[74] C.-C. Sun, Y.-Y. Chang, Y.-H. Wang, C.-Y. Chen, Y.-C. Lo, and H.-H. Cheng, "Precise Spatial Color Optical Modeling in Phosphor-Converted White LEDs", *Journal of Display Technology*, vol. 11, no. 3, pp. 261–265, 2015.

[75] A. A. Setlur, "Phosphors for LED-based solid-state lighting", *The Electrochemical Society Interface*, vol. 16, no. 4, pp. 32–36, 2009.

[76] H. Jin, S. Jin, K. Yuan, and S. Cen, "Two-Part Gauss Simulation of Phosphor-Coated LED", *IEEE Photonics Journal*, vol. 5, no. 4, p. 1 600 110, 2013.

[77] S. S. Jeong and J.-H. Ko, "Analysis of the spectral characteristics of white light-emitting diodes under various thermal environments", *Journal of Information Display*, vol. 13, no. 1, pp. 37–42, 2012.

[78] A. Vaskuri, H. Baumgartner, P. Kärhä, G. Andor, and E. Ikonen, "Modeling the spectral shape of InGaAlP-based red light-emitting diodes", *Journal of Applied Physics*, vol. 118, no. 20, p. 203 103, 2015.

[79] A. Thorseth, "Characterization, modeling, and optimization of light-emitting diode systems", PhD thesis, Technical University of Denmark, 2011.

[80] G. He and L. Zheng, "A model for LED spectra at different drive currents", *Chinese Optics Letters*, vol. 8, no. 11, pp. 1090–1094, 2010.

[81] F. Reifegerste and J. Lienig, "Modelling of the Temperature and Current Dependence of LED Spectra", *Journal of Light & Visual Environment*, vol. 32, no. 3, pp. 288–294, 2008.

[82] A. Keppens, W. Ryckaert, G. Deconinck, and P. Hanselaer, "Modeling high power light-emitting diode spectra and their variation with junction temperature", *Journal of Applied Physics*, vol. 108, no. 4, p. 043 104, 2010.

[83] I. Rotscholl, K. Trampert, U. Krüger, M. Perner, F. Schmidt, and C. Neumann, "Determination of tailored filter sets to create rayfiles including spatial and angular resolved spectral information", *Optics Express*, vol. 23, no. 23, pp. 29 543–29 554, 2015.

[84] Schott AG, *TIE-35: Transmittance of optical glass*, 2005, (accessed July 2015).

[85] Arnaud Darmont, Aphesa, *Spectral Response of Silicon Image Sensors*, White paper, 2009 (accessed July 2017).

[86] Z. M. Zhang, T. R. Gentile, A. L. Migdall, and R. U. Datla, "Transmittance measurements for filters of optical density between one and ten", *Applied Optics*, vol. 36, no. 34, pp. 8889–8895, 1997.

[87] Commission Internationale de l'Eclairage, *CIE Technical Report 202:2011: Spectral Responsivity Measurement of Detectors, Radiometers and Photometers*, 2011.

[88] H. M. Antia, *Numerical Methods for Scientists and Engineers*, 2. ed. Birkhäuser, 2002.

[89] Schott AG, *Optical Filters*, 2014, (accessed July 2015).

[90] CVI Laser Optics, *Datasheet selection guide*, pp. 655-658, (accessed July 2015).

[91] Oriel Instruments, *Filter characteristics*, pp. 10-30 - 10-32, (accessed July 2015).

[92] I. H. Blifford, "Factors Affecting the Performance of Commercial Interference Filters", *Applied Optics*, vol. 5, no. 1, pp. 105–111, 1966.

[93] Commission Internationale de l'Eclairage, *CIE Technical Report 198:2011: Determination of Measurement Uncertainties in Photometry*, 2011.

[94] U. Krüger and F. Schmidt, "The impact of cooling on CCD-based camera systems in the field of image luminance measuring devices", *Metrologia*, vol. 46, no. 4, S252, 2009.

[95] D. C. Montgomery, *Design and Analysis of Experiments*, 7. ed. Wiley, 2009.

[96] R. F. Gunst and R. L. Mason, "Fractional factorial design", *Wiley Interdisciplinary Reviews: Computational Statistics*, vol. 1, no. 2, pp. 234–244, 2009.

[97] Newport, *Newport resource e-catalog*, pp. 655-658, (accessed July 2015).

[98] I. Rotscholl, M. Katona, K. Trampert, U. Krüger, F. Schmidt, and C. Neumann, "Hyperspectral LED models for optical simulations", *Optics Express*, vol. 24, no. 26, A1597–A1611, 2016.

[99] J. R. W. Cieszyński and M. Wójcik, "Advanced light sources with rays for rendering engines", in *Proceedings of the International Light Simulation Symposium*, 2012, pp. 25–42.

[100] T.-C. Lu and C.-Y. Chang, "A Survey of VQ Codebook Generation", *Journal of Information Hiding and Multimedia Signal Processing*, vol. 1, no. 3, pp. 190–203, 2010.

[101] Markus Deserno, Max-Planck-Institut für Polymerforschung, *How to generate equidistributed points on the surface of a sphere*, White paper, 2004 (accessed September 2014).

[102] L. Lovisolo and E. Da Silva, "Uniform distribution of points on a hyper-sphere with applications to vector bit-plane encoding", *IEE Proceedings-Vision, Image and Signal Processing*, vol. 148, no. 3, pp. 187–193, 2001.

[103] M. Katona, I. Rotscholl, K. Trampert, and C. Neumann, "Übersicht über die Berechnung von Lichtstärkeverteilungen aus Rayfiles mit ihren Vor- und Nachteilen unter Verwendung verschiedener Raumwinkeltypen", in *Proceedings of the Lux junior*, 2017.

[104] R. L. Burden and J. D. Faires, *Numerical Analysis*, 9. ed. Brooks/Cole, 2011.

[105] S. Semiconductors, *Z-Power series Datasheet F50360*, 2008.

[106] I. Sayanca, K. Trampert, and C. Neumann, "Fehlanpassung der Normspektralwertfunktionen bei spektral integrierender Farbmessung", in *Proceedings of the Lux junior*, 2017.

[107] Y. J. Jung, H. Sohn, S.-i. Lee, Y. M. Ro, and H. W. Park, "Quantitative measurement of binocular color fusion limit for non-spectral colors", *Optics Express*, vol. 19, no. 8, pp. 7325–7338, 2011.

[108] J. Wang, Y. Zhang, X. Li, and Y. Weng, "Color breakup visibility thresholds for 2-field sequential colors", *Color Research & Application*, vol. 42, no. 5, pp. 580–590, 2017.

[109] Y. H. Kim, P. Arunkumar, B. Y. Kim, S. Unithrattil, E. Kim, S.-H. Moon, J. Y. Hyun, K. H. Kim, D. Lee, J.-S. Lee, *et al.*, "A zero-thermal-quenching phosphor", *Nature Materials*, vol. 16, no. 5, pp. 543–550, 2017.

[110] P. F. Smet and J. J. Joos, "White light-emitting diodes: Stabilizing colour and intensity", *Nature Materials*, vol. 16, no. 5, pp. 500–501, 2017.

[111] N. D. Q. Anh, N. T. P. Thao, and M. Voznak, "Green-emitting (La, Ce, Tb)PO_4:Ce:Tb phosphor: A novel solution for an increase in color homogeneity of white LEDs", in *6th International Symposium on Next Generation Electronics*, 2017, pp. 1–2.

APPENDIX A

ADDITIONAL INFORMATION REGARDING PMBS

A.1 INVARIANCE OF PMBS TOWARDS INITIAL AMPLITUDES OF MODELED BASIS SPECTRA

It is important to show that the PMBS method tolerates any individual non-zero arbitrary weighting factors p_n for each basis spectrum such that

$$S_n^{'}(\lambda) = p_n S_n(\lambda) \tag{A.1}$$

without affecting the final result, which is the reconstructed plenoptic function according to Eq. 4.12. The plenoptic function becomes

$$R(X, Y, Z, \phi, \vartheta, \lambda) = \sum_{n=1}^{n_{\max}} A_n(X, Y, Z, \phi, \vartheta) \times S_n(\lambda), \tag{A.2}$$

if the arbitrary factor p_n is not applied (first case) and

$$R^{'}(X, Y, Z, \phi, \vartheta, \lambda) = \sum_{n=1}^{n_{\max}} A_n^{'}(X, Y, Z, \phi, \vartheta) \times S_n^{'}(\lambda), \tag{A.3}$$

if the arbitrary factor p_n is applied (second case). To prove the invariance of the plenoptic function towards an initial basis spectrum weighting, it has to be shown that the term $R(X, Y, Z, \phi, \vartheta, \lambda)$ is equal to $R^{'}(X, Y, Z, \phi, \vartheta, \lambda)$. This is valid if $A_n^{'}$ is reconstructed with the inverse weighting factor p_n such that

$$A_n^{'}(X, Y, Z, \phi, \vartheta) = \frac{1}{p_n} A_n(X, Y, Z, \phi, \vartheta) \tag{A.4}$$

If the factor p_n is applied the matrix $M_{\mathrm{S}\tau}$ changes to

$$M'_{\mathrm{S}\tau} = \begin{bmatrix} \int p_1 S_1(\lambda)\tau_{\mathrm{eff},1}(\lambda)d\lambda & \cdots & \int p_{n_{max}} S_{n_{max}}(\lambda)\tau_{\mathrm{eff},1}(\lambda)d\lambda \\ \vdots & \ddots & \vdots \\ \int p_1 S_1(\lambda)\tau_{\mathrm{eff},n_{max}}(\lambda)d\lambda & \cdots & \int p_{n_{max}} S_{n_{max}}(\lambda)\tau_{\mathrm{eff},n_{max}}(\lambda)d\lambda \end{bmatrix}$$
$$= M_{\mathrm{S}\tau} \begin{bmatrix} p_1 & \cdots & 0 \\ \vdots & \ddots & \vdots \\ 0 & \cdots & p_n \end{bmatrix} \tag{A.5}$$

The reconstruction equation in the first case is given as

$$\begin{bmatrix} A_1(X,Y,Z,\phi,\vartheta) \\ \vdots \\ A_{n_{\mathrm{max}}}(X,Y,Z,\phi,\vartheta) \end{bmatrix} = M_{\mathrm{S}\tau}^{-1} \cdot \begin{bmatrix} M_{\mathrm{A},1}(X,Y,Z,\phi,\vartheta) \\ \vdots \\ M_{\mathrm{A},n_{\mathrm{max}}}(X,Y,Z,\phi,\vartheta) \end{bmatrix} \tag{A.6}$$

and in the second case as

$$\begin{bmatrix} A'_1(X,Y,Z,\phi,\vartheta) \\ \vdots \\ A'_{n_{max}}(X,Y,Z,\phi,\vartheta) \end{bmatrix} = M'^{-1}_{\mathrm{S}\tau} \cdot \begin{bmatrix} M_{\mathrm{A},1}(X,Y,Z,\phi,\vartheta) \\ \vdots \\ M_{\mathrm{A},n_{max}}(X,Y,Z,\phi,\vartheta) \end{bmatrix}$$

$$\begin{bmatrix} A'_1(X,Y,Z,\phi,\vartheta) \\ \vdots \\ A'_{n_{max}}(X,Y,Z,\phi,\vartheta) \end{bmatrix} = \begin{bmatrix} p_1 & \cdots & 0 \\ \vdots & \ddots & \vdots \\ 0 & \cdots & p_{n_{max}} \end{bmatrix}^{-1} M_{\mathrm{S}\tau}^{-1} \begin{bmatrix} M_{\mathrm{A},1}(X,Y,Z,\phi,\vartheta) \\ \vdots \\ M_{\mathrm{A},n_{max}}(X,Y,Z,\phi,\vartheta) \end{bmatrix} \tag{A.7}$$

or

$$\begin{bmatrix} p_1 & \cdots & 0 \\ \vdots & \ddots & \vdots \\ 0 & \cdots & p_{n_{max}} \end{bmatrix} \begin{bmatrix} A'_1(X,Y,Z,\phi,\vartheta) \\ \vdots \\ A'_{n_{max}}(X,Y,Z,\phi,\vartheta) \end{bmatrix} = M_{\mathrm{S}\tau}^{-1} \begin{bmatrix} M_{\mathrm{A},1}(X,Y,Z,\phi,\vartheta) \\ \vdots \\ M_{\mathrm{A},n_{max}}(X,Y,Z,\phi,\vartheta) \end{bmatrix}. \tag{A.8}$$

The right hand side of Eq. A.6 is the same as A.8 and it can therefore be written that

$$\begin{bmatrix} p_1 & \cdots & 0 \\ \vdots & \ddots & \vdots \\ 0 & \cdots & p_{n_{max}} \end{bmatrix} \begin{bmatrix} A_1'(X,Y,Z,\phi,\vartheta) \\ \vdots \\ A_{n_{max}}'(X,Y,Z,\phi,\vartheta) \end{bmatrix} = \begin{bmatrix} A_1(X,Y,Z,\phi,\vartheta) \\ \vdots \\ A_{n_{max}}(X,Y,Z,\phi,\vartheta) \end{bmatrix}, \tag{A.9}$$

which shows that Eq. A.4 is fulfilled.

Hence, a fundamental difference exists between the influence of the arbitrary basis spectra amplitude matrix and the measurement weighting matrix, which are both inserted into the reconstruction equation:

$$\begin{bmatrix} W_{R,1} & \cdots & 0 \\ \vdots & \ddots & \vdots \\ 0 & \cdots & W_{R,n_{max}} \end{bmatrix} \begin{bmatrix} M_{R,1} \\ \vdots \\ M_{R,n_{max}} \end{bmatrix} = M_{S\tau} \cdot \begin{bmatrix} p_1 & \cdots & 0 \\ \vdots & \ddots & \vdots \\ 0 & \cdots & p_{n_{max}} \end{bmatrix} \cdot \begin{bmatrix} A_1 \\ \vdots \\ A_{n_{max}} \end{bmatrix} \tag{A.10}$$

or

$$\begin{bmatrix} A_1 \\ \vdots \\ A_{n_{max}} \end{bmatrix} = \begin{bmatrix} \frac{1}{p_1} & \cdots & 0 \\ \vdots & \ddots & \vdots \\ 0 & \cdots & \frac{1}{p_{n_{max}}} \end{bmatrix} M_{S\tau}^{-1} \begin{bmatrix} W_{R,1} & \cdots & 0 \\ \vdots & \ddots & \vdots \\ 0 & \cdots & W_{R,n_{max}} \end{bmatrix} \begin{bmatrix} M_{R,1} \\ \vdots \\ M_{R,n_{max}} \end{bmatrix}. \tag{A.11}$$

This is

$$\begin{aligned} A_1(X,Y,Z,\phi,\vartheta) &= \frac{W_A}{p_1} \times \Big[M_1(X,Y,Z,\phi,\vartheta) \times M_{S\tau}^{-1}(1,1) W_{R,1} \\ &\quad + \cdots \\ &\quad + M_{n_{max}}(X,Y,Z,\phi,\vartheta) \times M_{S\tau}^{-1}(1,n_{max}) W_{R,n_{max}} \Big] \\ &\ \ \vdots \\ A_{n_{max}}(X,Y,Z,\phi,\vartheta) &= \frac{W_A}{p_{n_{max}}} \times \Big[M_1(X,Y,Z,\phi,\vartheta) \times M_{S\tau}^{-1}(n_{max}1) W_{R,1} \\ &\quad + \cdots \\ &\quad + M_{n_{max}}(X,Y,Z,\phi,\vartheta) \times M_{S\tau}^{-1}(n_{max},n_{max}) W_{R,n_{max}} \Big] \end{aligned} \tag{A.12}$$

The reason for the different influences of these two diagonal matrices is that the basis spectra weighting including the p_n is multiplied from the left hand side and the measurement weighting matrix including the relative weightings $W_{R,n}$ is multiplied from the right hand side to the matrix $M_{S\tau}^{-1}$. Therefore the correct determination of the measurement weighting matrix is crucial while the basis spectra weighting can be arbitrary as long as it is non-zero.

A.2 DETAILED EXPLANATION OF SPATIAL SEPARATION VALUE $P_{SUM,n}$

The figure of merit used to quantify the spatial separation $P_{sum,n}$ of the nth individual spectral model according to section 5.3 is explained in detail below. The starting point is a reconstructed radiance image (or irradiance of a projected ray file) $A(X,Y)$, which is normalized to its integral such that $\sum_X \sum_Y A(X,Y) = 1$. Since a high spatial separation and localization consists of few high values and as many zeros (or dark areas) as possible, the total number of pixels above a threshold Th defined as

$$P_{tot}(Th) = \sum_X \sum_Y P(X,Y,Th) \tag{A.13}$$

with

$$P(X,Y,Th) = \begin{cases} 0 & |A(X,Y)| \leq Th \\ 1 & |A(X,Y)| > Th \end{cases} \tag{A.14}$$

can be used as an indicator. $P_{tot}(Th)$ is a function of the threshold value and the threshold value lies in the interval $Th \in (0\,1)$ because all values $|A(X,Y)|$ have to lie between 0 and 1. Since $P_{tot}(Th) \geq 0$ $\forall\ Th$, the cumulative sum $P_{cumsum}(Th)$ defined as

$$P_{cumsum}(Th) = \int_{j=0}^{Th} P_{tot}(j) \tag{A.15}$$

is a monotone increasing function. Furthermore, it holds that $P_{\text{tot}}(Th)$ itself is a monotone decreasing function. Another important attribute of $P_{\text{cumsum}}(Th)$ is that its maximum value $P_{\text{cumsum}}(Th = 1) = const.$ due to the normalization $\sum_X \sum_Y A(X,Y) = 1$. The figure of merit for the spatial separation/narrow localization is then defined as

$$P_{\text{sum},n} = \int^{Th} P_{\text{cumsum}}(Th)dTh. \tag{A.16}$$

Figure A.1 shows three different artificially reconstructed 1-dimensional amplitudes $A(X)$ with different spatial separations/narrow localizations. They are normalized to their integrals.

"Spatial 1" shows data localized evenly on two pixels ($X = 3$ and $X = 5$). "Spatial 2" is localized more strongly at pixel $X = 5$ and more weakly at pixel $X = 3$. "Spatial 3" is localized completely at pixel $X = 5$. The localization decreases from Spatial 1 to Spatial 3. Note that in the case of the PMBS basis spectra validation, each tested distribution would represent a different reconstruction (based on a different set of basis spectra).

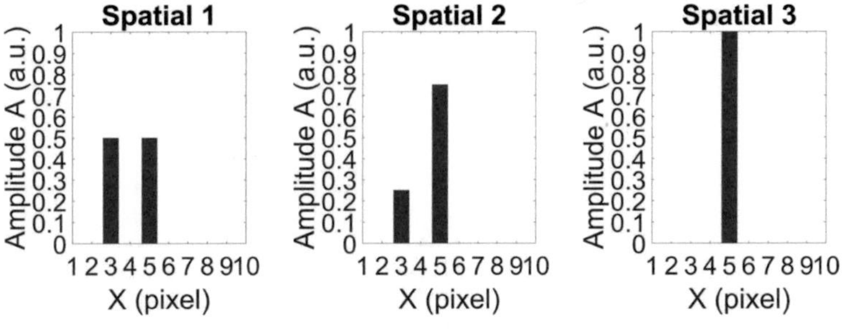

(a) Spatial 1 **(b)** Spatial 2 **(c)** Spatial 3

Figure A.1: Artificially reconstructed 1-dimensional amplitudes $A(X)$ with different spatial separations/narrow localizations.

Figure A.2 (a)-(c) show the $P_{\text{tot}}(Th)$ for Spatial 1 to Spatial 3. In the case of Spatial 3, $P_{\text{tot}}(Th) = 1$ for each threshold value. In the case of Spatial 2, $P_{\text{tot}}(Th) = 2$ until $Th = A(X = 3)$. Then $P_{\text{tot}}(Th) = 1$ because only pixel $X = 5$ is above the threshold. After reaching the value $Th = A(X = 5)$, $P_{\text{tot}}(Th)$ drops to zero. Spatial 1 is two as pixels $X = 3$ and $X = 5$ are above all thresholds until the threshold reaches $Th = A(X = 3) = A(X = 5)$. Subsequently, $P_{\text{tot}}(Th) = 0$. Figure A.2 (d) shows the resulting $P_{\text{cumsum}}(Th)$ for all three distributions. It can be seen that all reach the same maximum value. It is 200 since the threshold value was increased evenly 200 times between 0 and 1 in this example. The important point is that it not depends on the normalized $A(X)$. Figure A.2 (e) displays the integrals P_{sum} used as figure of merit for the spatial separation/narrow localization. As expected, Spatial 3, which is localized perfectly has the lowest value, while Spatial 1, which is localized least perfectly has the highest value.

$P_{\text{tot}}(Th)$ always decreases with an increasing threshold Th. A low threshold results in a high P_{tot} because many pixels near zero are above the threshold. If Th is increased, it starts to include less pixels and therefore decreases. $P_{\text{cumsum}}(Th)$ always increases (or stays constant). The maximum value depends only on the number of discrete values used for the threshold value Th as described above.

In the case of a low spatial separation the first values of $P_{\text{cumsum}}(Th)$ are higher as is the slope of $P_{\text{cumsum}}(Th)$. If fewer pixels are above the threshold, the slope gets smaller and may reach zero eventually. The important fact is that the value $P_{\text{cumsum}}(Th)$ of a high spatial separation is larger or equal to $P_{\text{cumsum}}(Th)$ of a low spatial separation for each threshold value. This is always true because $P_{\text{cumsum}}(Th)$ are monotone increasing functions for all $A(X, Y)$ with the same maximal value and both derivations $P_{\text{tot}}(Th)$ are decreasing functions. Therefore, the integral $P_{\text{sum},n} = \int^{Th} P_{\text{cumsum}}(Th)dTh$ can be used as a direct measure of the spatial separation of the nth spectral model.

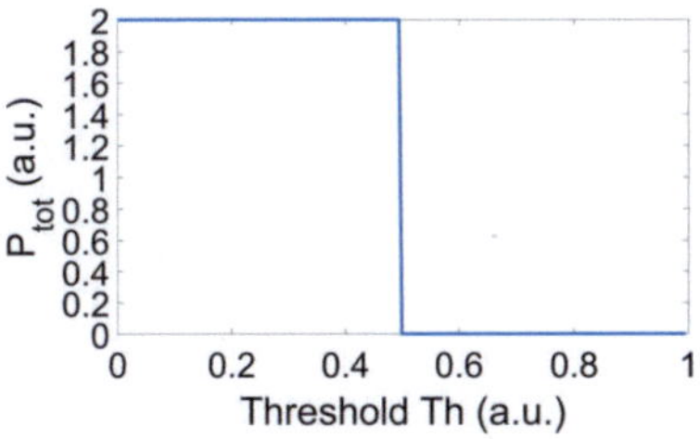

(a) $P_{\mathrm{tot}}(Th)$ of Spatial 1

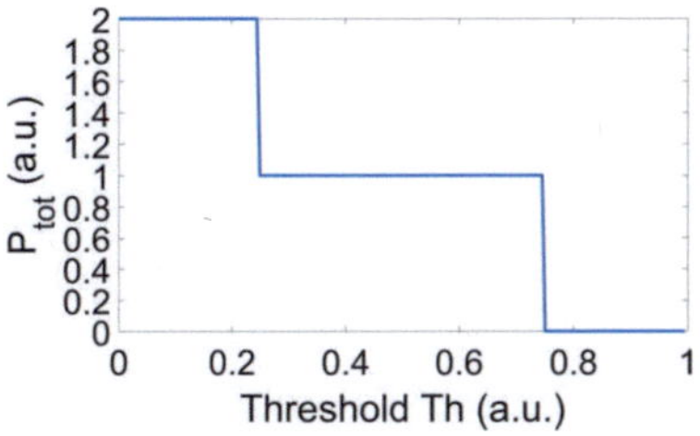

(b) $P_{\mathrm{tot}}(Th)$ of Spatial 2

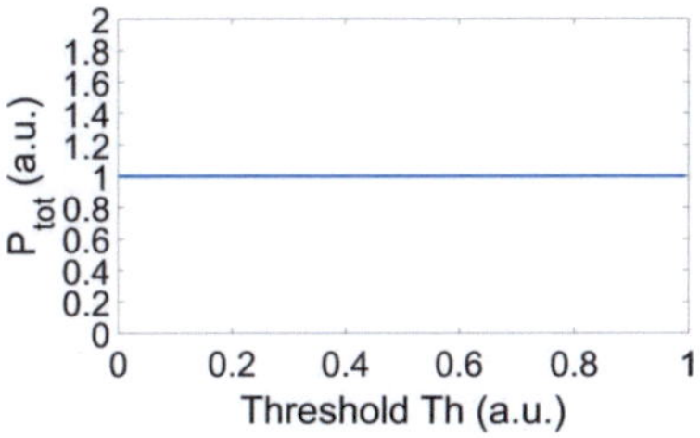

(c) $P_{\mathrm{tot}}(Th)$ of Spatial 3

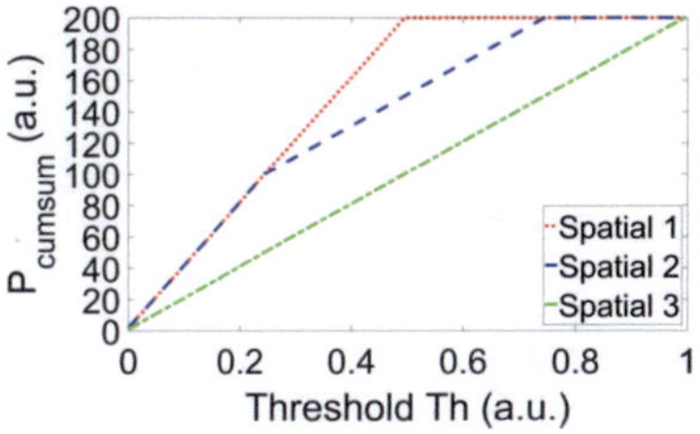

(d) $P_{\mathrm{cumsum}}(Th)$ for Spatial 1-3

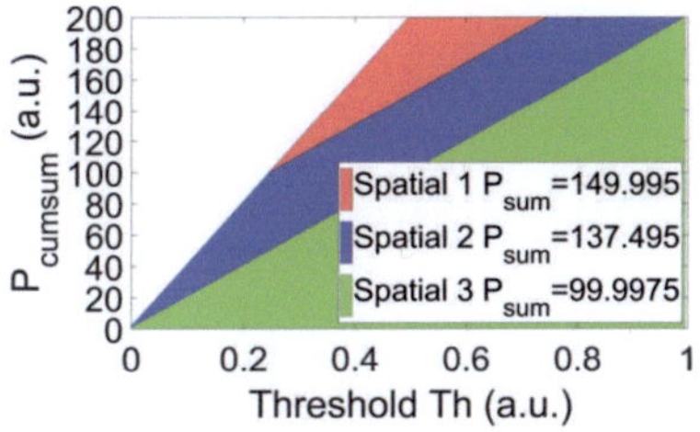

(e) P_{sum} for Spatial 1-3

Figure A.2: $P_{\mathrm{tot}}(Th)$, $P_{\mathrm{cumsum}}(Th)$ and P_{sum} for Spatial 1-3.

Appendix B

Spectral ray tracing alignment consideration

The quantitative evaluation of the spectral ray tracing in section 8.3 did not consider the deviations at the edges of the simulated spectral distribution. Furthermore, it considered only the three major sequences of the ray tracing, which contain 92% of the radiometric power. The remaining sequences were ignored. To justify both steps it has to be shown that neither the stray light nor the deviations are caused by the spectral ray file but occur due to misalignments between the near field measurement and the simulation, misalignments between the far field measurement and the simulation, or deviating material parameters in the simulation. As the problem is identical for the spectral ray tracing of the Red/White example and the RGBW example, it is adequate to consider only one of them below.

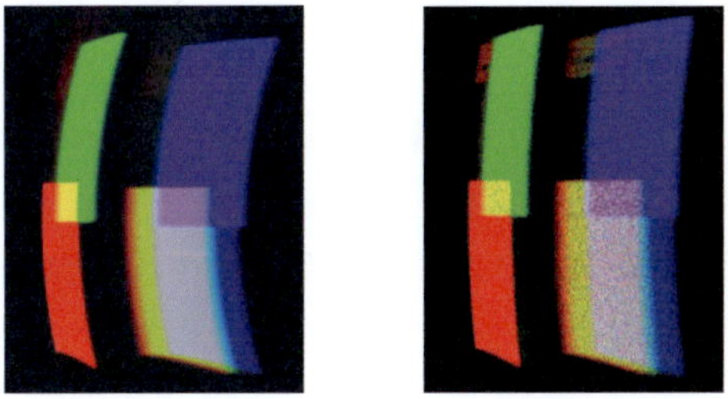

(a) (b)

Figure B.1: Comparison of measurement (a) and spectral ray tracing (b) with Red/Green/Blue/White ray file from section 8.3.

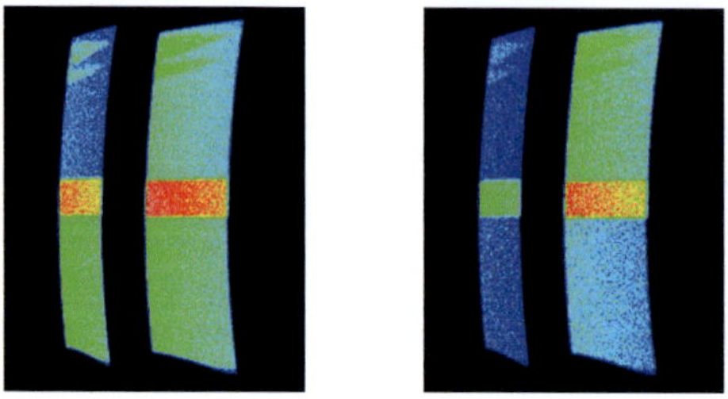

(a) (b)

Figure B.2: Comparison of shape obtained from ray tracing using originally measured ray files (a) and ray files from PMBS (b).

Figure B.1 shows the measured and simulated distributions from the RGBW example of section 8.3. The two main deviations regarding the shape are the curvature at the edges on the right hand side and the relative positions of the green and blue distribution. The green distribution is located above the blue distribution only in the measurement.

To show that these deviations are not caused by the spectral ray files, a simulation with original measured ray files is compared to a simulation with spectral ray files. As no spectrum can be assigned to the original ray files, all wavelengths were set to 555 nm. Figure B.2 shows the normalized intensity obtained with a simulation using the original ILMD-based ray files. The ray files are based on the standard filters $x_{\text{short}}(\lambda)$, $x_{\text{long}}(\lambda)$ and $v(\lambda)$ because they provide a sufficient signal for all sources. Figure B.2 shows the obtained normalized intensity of a simulation with the spectral ray files from the blue, green, red and phosphor model obtained with PMBS from section 8.3. The comparison of both distributions show that both the shape and relative position of the green and blue distributions are identical for both ray file sets. Therefore these deviations are not caused by the spectral ray files.

The same comparison can also be used to verify whether the 8% stray light only occur in the simulation using the spectral ray file set. Figure B.3 (a) and (b) show the unfiltered simulation with the stray light on a logarithmic scale. As the stray light distributions are comparable, it shows that they are not caused by the spectral ray files either. Note that the relative deviations in all distributions are expected because the weighting of the individual sources is different in the original ray files compared to those of the spectral ray files.

(a) **(b)**

Figure B.3: Comparison of stray light obtained from ray tracing using originally measured ray files (a) and ray files from PMBS (b) on a logarithmic scale.

List of Symbols

Symbol	Description
α_n	sum of all negative elements in nth row of the reconstruction matrix
β_n	sum of all positive elements in nth row of the reconstruction matrix
γ	angle of incidence
ΔA	vector of $n_{\max}$ ΔA_n corresponding to a set of ILMD-based measurements
ΔA_n	value of $\Delta A_n(X, Y, Z, \phi, \vartheta)$
$\Delta A_n(X, Y, Z, \phi, \vartheta)$	deviations of $A_n(X, Y, Z, \phi, \vartheta)$ from true amplitude distribution
$\Delta u'v'$	chromaticity distance in CIE $u'v'$ diagram
$\Delta u'v'_{\text{median}}$	median of several $\Delta u'v'$
$\Delta u'v'_{\text{mean}}$	mean value of several $\Delta u'v'$
ϵ	precision of the false position or bisection
ζ	calculation success of spectral reconstruction
η	vector of $n_{\max}$ η_n corresponding to a set of ILMD-based measurements
$\eta_{\tau,n}$	uncertainties regarding specific $s_{\tau,\text{eff},n}(\lambda)$
η_{N}	uncertainties affecting all $M_n(X, Y, Z, \phi, \vartheta)$
η_n	value of $\eta_n(X, Y, Z, \phi, \vartheta)$

Symbol	Description
$\eta_n(X,Y,Z,\phi,\vartheta)$	deviations of $M_n(X,Y,Z,\phi,\vartheta)$ to $M_{t,n}(X,Y,Z,\phi,\vartheta)$
ϑ	polar angle of radiation
ϑ_S	polar angle of starting points X, Y and Z on surface
λ	wavelength
λ_C	peak wavelength of spectral model
λ_{Cut} ($\lambda_{Cut,n}$)	(n^{th}) separation wavelength
$\lambda_{i,C}$	peak wavelength of ith spectral model
ν	frequency
ν_C	peak frequency of spectral model
$\sigma_{i,n}$	n^{th} parameter for width and skewness of ith spectral model
σ_n	n^{th} parameter for width and skewness of spectral model
$\tau(\lambda)$	optical transmission
$\tau_{Fr}(\gamma, n_r)$	transmission according to Fresnel equations
$\tau_{in}(\lambda)$	internal transmission spectrum
$\tau_n(\lambda)$	$\tau(\lambda)$ of used filter in n^{th} ILMD-based measurement
$\tau_{ND,n}(\lambda)$	transmission of neutral density used in n^{th} ILMD-based measurement
ϕ	azimuth angle of radiation
ϕ_S	azimuth angle of starting points X, Y and Z on surface
A	vector of n_{max} A_n corresponding to a set of ILMD-based measurements
$A(X)$	artificial one-dimensional $A(X,Y)$
$A(X,Y)$	radiance image or irradiance based on a basis spectrum

Symbol	Description
$A(X,Y,Z,\phi,\vartheta)$	radiometric non-spectral and non-temporal plenoptic function of individual basis spectrum
$A_{\eta,n}$	$A_n(X,Y,Z,\phi,\vartheta)$ including simulated ΔA_n
$A_{angular}(Voron(\phi_S,\vartheta_S),\phi,\vartheta)$	angular histogram of spectral ray file
$A_{angular,n}(Voron(\phi_S,\vartheta_S),\phi,\vartheta)$	angular histogram of nth spectral ray file
$A_{\text{Blue}}(X,Y,Z,\phi,\vartheta)$	$A(X,Y,Z,\phi,\vartheta)$ of blue LED
A_n	values of $A_{\text{spatial},n}(\phi_{\text{S}},\vartheta_{\text{S}})$ or $A_{\text{angular},n}(Voron(\phi_S,\vartheta_S),\phi,\vartheta)$ or $A(X,Y,Z,\phi,\vartheta)$
A_n^- / A_n^+	negative/positive A_n
$A_n(\phi,\vartheta)$	specific spectral histogram of nth spectral ray file
$A_n(\phi,\vartheta,r)$	amplitude of nth basis spectrum at known position (ϕ,ϑ,r)
$A_n(X,Y)$	radiance image or irradiance based on nth basis spectrum
$A_n(X,Y,Z,\phi,\vartheta)$	nth $A(X,Y,Z,\phi,\vartheta)$
$A_{\text{Phosphor}}(X,Y,Z,\phi,\vartheta)$	$A(X,Y,Z,\phi,\vartheta)$ of phosphor
$A_{\text{spatial}}(\phi_{\text{S}},\vartheta_{\text{S}})$	spatial histogram of spectral ray file
$A_{\text{spatial},n}(\phi_{\text{S}},\vartheta_{\text{S}})$	spatial histogram of nth spectral ray file
ADC n	ADC of nth voronoi region
d	thickness of optical filter
d_0	reference thickness
$d\lambda$	wavelength increment
$d\lambda/dT$	temperature coefficient of λ

Symbol	Description
$f(x)$	function of x
$I_{\text{mean},n,r}$	mean value of $I_{n,r}(\phi,\vartheta)$
$I_{n,r}(\phi,\vartheta)$	pseudo-intensity of ray file created with $M(\phi_S,\vartheta_S,\phi,\vartheta)$
$J_{\text{mean},n,r}$	mean value of $J_{n,r}(\phi,\vartheta)$
$J_{n,r}(\phi,\vartheta)$	pseudo-intensity of measured ray file (original ray file)
$k_i(\lambda)$	i^{th}arbitrary weighting function for RSS merit
M	vector of n_{max} M_n corresponding to a set of ILMD-based measurements
$M(\phi_S,\vartheta_S,\phi,\vartheta)$	model of plenoptic function created with measured ray file
$M(X,Y)$	pseudo-radiance image or pseudo-irradiance based on ray file of ILMD-based measurement
$M(X,Y,Z,\phi,\vartheta)$	unweighted pseudo-radiometric or photometric non-spectral and non-temporal plenoptic function
$M_{\eta,n}$	value of $M_{\eta,n}(X,Y,Z,\phi,\vartheta)$
$M_{\eta,n}(X,Y,Z,\phi,\vartheta)$	$M_n(X,Y,Z,\phi,\vartheta)$ including simulated $\eta_n(X,Y,Z,\phi\vartheta)$
$M_{\text{A}}(X,Y,Z,\phi,\vartheta)$	$M(X,Y,Z,\phi,\vartheta)$ with absolute weighting
$M_{\text{A},m}(X,Y,Z,\phi,\vartheta)$	m^{th} $M(X,Y,Z,\phi,\vartheta)$ with absolute weighting
$M_{\text{A},n}(\phi,\vartheta,r)$	absolute measurement value of nth ILMD-based measurement at known position (ϕ,ϑ,r)
$M_{\text{A},n}(X,Y,Z,\phi,\vartheta)$	n^{th} $M(X,Y,Z,\phi,\vartheta)$ with absolute weighting

Symbol	**Description**
$M_{\mathrm{A,SB}}(X,Y,Z,\phi,\vartheta)$	sharp blue $M(X,Y,Z,\phi,\vartheta)$ with absolute weighting
$M_{\mathrm{A,SY}}(X,Y,Z,\phi,\vartheta)$	sharp yellow $M(X,Y,Z,\phi,\vartheta)$ with absolute weighting
$M_{angular}(ADC\,n,\phi,\vartheta)$	angular histogram of nth specific voronoi region of measured ray file
$M_{angular}(Voron(\phi_S,\vartheta_S),\phi,\vartheta)$	angular histogram of measured ray file
$M_{angular,n}(Voron(\phi_S,\vartheta_S),\phi,\vartheta)$	angular histogram of nth measured ray file
$M_{\mathrm{min},n}, M_{\mathrm{max},n}, M_{\mathrm{mean},n}$	obtained minimal, maximal and mean $M_{\eta,n}$ used in factorial design
M_n	value of $M_n(X,Y,Z,\phi,\vartheta)$, pseudo-radiometric or photometric value
$M_n(\phi,\vartheta)$	specific measurement histogram of nth measured ray file
$M_n(\phi,\vartheta,r)$	counted ray number of nth measured ray file at known position (ϕ,ϑ,r)
$M_n(\phi_S,\vartheta_S,\phi,\vartheta)$	model of plenoptic function created with n^{th} measured ray file
$M_n(X,Y)$	pseudo-radiance image or pseudo-irradiance based on ray file of nth ILMD-based measurement
$M_n(X,Y,Z,\phi,\vartheta)$	n^{th} $M(X,Y,Z,\phi,\vartheta)$
$M_{\mathrm{R,S}\tau}$	normalized $M_{\mathrm{S}\tau}$
$M_{\mathrm{S}\tau}^{-1}$ $(M_{\mathrm{S}\tau})$	(inverse) spectral reconstruction matrix
$M_{\mathrm{S}\tau,\mathrm{RGB}}^{-1}$	spectral reconstruction matrix of RGB LED

Symbol	Description
$M_{S,n}(X,Y,Z,\phi,\vartheta)$	nth sharp $M(X,Y,Z,\phi,\vartheta)$
$M_{SB}(X,Y,Z,\phi,\vartheta)$	sharp blue $M(X,Y,Z,\phi,\vartheta)$
$M_{spatial}(\phi_S,\vartheta_S)$	spatial histogram of measured ray file
$M_{spatial,n}(\phi_S,\vartheta_S)$	spatial histogram of nth measured ray file
$M_{SY}(X,Y,Z,\phi,\vartheta)$	sharp yellow $M(X,Y,Z,\phi,\vartheta)$
$M_{t,n}$	value of $M_n(X,Y,Z,\phi,\vartheta)$
$M_{t,n}(X,Y,Z,\phi,\vartheta)$	$M_n(X,Y,Z,\phi,\vartheta)$ without any uncertainty
n, m	index variables, number
$N_{angular}$	number of spatial histogram bins used to sample ray files
N_{comb}	number of possible filter combinations
n_{eff}	effective index of refraction (used for interference filter)
n_{max}	number of basis spectra
n_r	refractive index
N_{PS}	selection number of filter preselection
N_{Ray}	total numbers of rays in ray file
$N_{Ray,n}$	total numbers of rays in the nth ray file
N_{Se}	selection number of filter selection
$N_{spatial}$	number of spatial histogram bins used to sample ray files
N_{Voron}	number of voronoi regions with constant ADC used to sample ray files
NCC	normalized cross-correlation
$NCC_{n,r}$	NCC of nth model at distance r
p	initial amplitude of spectral model
$P_{cumsum}(Th)$	cumulative sum of $P_{tot}(Th)$
p_i (p_n)	initial amplitude of ith (nth) spectral model
$P(X,Y,Z,\phi,\vartheta,\lambda,t)$	plenoptic function

Symbol	Description
$P_{\mathrm{Ref},n}$	reference spatial separation merit of $A_n(X,Y)$
p_{smooth}	smoothing parameter of smoothed spline
P_{sum}	spatial separation merit
$P_{\mathrm{sum},n}$	spatial separation merit of $A_n(X,Y)$
$P_{\mathrm{tot}}(Th)$	total number of pixels above Th
$Q_{\mathrm{aim},n}$	aim value of $Q_n(A_n)$ in optimization
$Q_{\mathrm{current},n}$	current value of $Q_n(A_n)$ in optimization
$Q_{\mathrm{final},n}$	aim value of $Q_n(A_n)$ in nearest neighbor calculation
$Q_n(\zeta)$	amount of negative/false positive values in A_n according to ζ
$Q_n(A_n)$	amount of negative/false positive values in A_n
$Q_{\mathrm{start},n}$	value of $Q_n(A_n)$ before nearest neighbor calculation
r	distance to point of origin of goniophotometer
$R(X,Y,Z,\phi,\vartheta,\lambda)$	non-temporal plenoptic function
$R_{\mathrm{S},n}(X,Y,Z,\phi,\vartheta,\lambda)$	n^{th} sharp $R(X,Y,Z,\phi,\vartheta,\lambda)$
$R_{\mathrm{SB}}(X,Y,Z,\phi,\vartheta,\lambda)$	sharp blue $R(X,Y,Z,\phi,\vartheta,\lambda)$
$R_{\mathrm{SY}}(X,Y,Z,\phi,\vartheta,\lambda)$	sharp yellow $R(X,Y,Z,\phi,\vartheta,\lambda)$
RSS	residual sum of squares
RSS_{G}	reference RSS (of Gaussian model)
RSS_{mean}	mean value of several RSS
$S(\lambda)$, $S(\nu)$	spectral power distribution (spectrum)
$S(\lambda,x)$	model spectrum of LED
$S_\eta(\lambda)$	reconstructed spectrum including simulated ΔA_n

Symbol	Description
$s_{\tau,\mathrm{eff}}(\lambda)$	effective spectral sensitivity of ILMD-system
$s_{\tau,\mathrm{eff},m}(\lambda)$	effective spectral sensitivity of ILMD-system used in mth ILMD-based measurement
$s_{\tau,\mathrm{eff},n}(\lambda)$	effective spectral sensitivity of ILMD-system used in nth ILMD-based measurement
$s_{\tau,n}(\lambda)$	$s_{\tau,\mathrm{eff},n}(\lambda)$ excluding $\tau_{\mathrm{ND},n}(\lambda)$
$S_{\mathrm{Blue}}(\lambda)$	blue LED spectrum
$S_{\mathrm{M}}(\lambda)$	measured spectrum
$S_{\mathrm{M}}(\lambda,\phi,\vartheta,r)$	measured spectrum at known position (ϕ,ϑ,r)
$S_n(\lambda)$	nth basis spectrum
$S_n(\lambda,x)$	nthmodel spectrum of LED
$S_{\mathrm{P}}(\lambda)$	spectrum modeled as smoothed spline (model of phosphor basis spectrum)
$S_{\mathrm{Phosphor}}(\lambda)$	phosphor spectrum
$S_{\mathrm{R,LED}}(\lambda)$	sum of remainder LED spectra
$S_{\mathrm{R,P}}(\lambda)$	remainder phosphor spectrum
$S_{S,n}(\lambda)$	nth sharp separated spectrum
$S_{\mathrm{SB}}(\lambda)$	sharp blue spectrum
$S_{\mathrm{SY}}(\lambda)$	sharp yellow spectrum
$s_{\mathrm{Sys}}(\lambda)$	spectral system sensivity
T	room temperature
t	time
T_{abs}	temperature drift caused by absorption
Th	threshold value in calculation of P_{sum}
$u(...)$	modeled uncertainty distribution of ...variable
$u(\lambda_\gamma)$	wavelength uncertainty induced by $u(\gamma)$
u'_n, v'_n	nth coordinates of CIE $u'v'$ diagram
u_σ	standard deviation of $u(...)$

Symbol	Description
$u_{\text{min}}, u_{\text{max}}, u_{\text{mean}}$	minimal, maximal and mean uncertainty $u(...)$ used in factorial design
$v(\lambda)$ $[= \overline{y}(\lambda)]$	photometric sensitivity of the human eye according to CIE
$Voron(\phi_S, \vartheta_S)$	specific voronoi region
W_{A}	absolute scaling factor
W_n	absolute weighting of nth ILMD-based measurement
$W_{\text{norm}}(W_{\text{norm},n})$	matrix to normalize $M_{S\tau}$ (nth diagonal element)
W_{R}	relative weighting matrix
$W_{\text{R},n}$	relative weighting factor of nth ILMD-based measurement
$W_{\text{S},n}$	nth global sharp weighting factor
W_{SB}	global sharp blue weighting factor
W_{SY}	global sharp yellow weighting factor
X, Y and Z	spatial X,Y and Z-dimension (starting point on a surface)
x	set of model parameters
$\overline{x}(\lambda), \overline{z}(\lambda), \overline{y}(\lambda)$	color matching functions according to CIE 1931
x_i	set of model parameters for ith basis LED spectrum
$X_{\text{CIE}}, Y_{\text{CIE}}, Z_{\text{CIE}}$	tristimulus values according to CIE 1931
$x_{\text{long}}(\lambda)$	filter functions which approximates $\overline{x}(\lambda)$ in the interval [500 830]
$x_{\text{short}}(\lambda)$	filter functions which approximates $\overline{x}(\lambda)$ in the interval [380 500]

Abbreviation	Description
ADC	angular distribution curve
BBY	Blue/Blue/Yellow (Phosphor)
BY	Blue/Yellow (Phosphor)
CAD	computer - aided design
CCD	charge-coupled device
CIE	Commission Internationale de l'Eclairage
CIE $u'v'$ diagram	CIE 1976 uniform chromaticity scale
CMOS	complementary metal-oxide-semiconductor
CRI	color rendering index
DUT	device under test
FWHM	full width at half maximum
GBY	Green/Blue/Yellow (Phosphor)
GmbH	german abbreviation for Gesellschaft mit beschränkter Haftung
IES	Illuminating Engineering Society
ILMD	imaging luminance measurement device
LED	light emitting diode
ND	neutral density
OLED	organic light emitting diode
PCA	principal component analysis
PDF	probability density function
PMBS	physically motivated basis spectra
RBY	Red/Blue/Yellow (Phosphor)
RGB	Red/Green/Blue
RGBW	Red/Green/Blue/White
RGBY	Red/Green/Blue/Yellow (Phosphor)
TM	Technical Memorandum

Publications by the Author

Journal papers

- I. Rotscholl, M. Katona, K. Trampert, U. Krüger, F. Schmidt and C. Neumann, "Hyperspectral LED models for optical simulations," *Optics Express*, vol. 24, no. 16, pp. A1597-A1611, 2016.
- I. Rotscholl, K. Trampert, U. Krüger, M. Perner, F. Schmidt and C. Neumann, "Determination of tailored filter sets to create rayfiles including spatial and angular resolved spectral information," *Optics Express*, vol. 23, no. 23, pp. 29543-29554, 2015.
- I. Rotscholl, K. Trampert, F. Herrmann and C. Neumann, "Methods for describing illumination colour uniformities," *Advanced Optical Technologies*, vol. 4, no. 1, pp. 99-110, 2015.
- I. Rotscholl, K. Trampert, U. Krüger, I. L. Sayanca, F. Schmidt and C. Neumann, "Measurement of angular and spatial resolved spectral rayfiles with conventional nearfield goniophotometers and standard optical filters," *LED Professional Review*, no. 63, pp. 48-56, 2017.

Some passages of this work have been quoted verbatim from the first two publications listed above.

CONFERENCE CONTRIBUTIONS

- I. Rotscholl, K. Trampert, U. Krüger, I. L. Sayanca, F. Schmidt and C. Neumann, "Measurement of angular and spatial resolved spectral rayfiles with conventional nearfield goniophotometers and standard optical filters," in *LED Professional Symposium*, 2016.
- I. Rotscholl, K. Trampert, U. Krüger, M. Perner, I. L. Sayanca, F. Schmidt and C. Neumann, "Bestimmung von winkel- und ortsaufgelösten spektralen Rayfiles," in *Licht*, 2016.
- I. Rotscholl and C. Neumann, "Die Farbtemperatur in der Lichttechnik," in *Proceedings of the Lux junior*, 2015.
- I. Rotscholl, K. Trampert and C. Neumann, "Evaluierung der Farbhomogenität von Lichtquellen," in *Licht- und Lebensqualität*, 2015.
- M. Katona, I. Rotscholl, K. Trampert and C. Neumann, "Determination of errors associated with the far field assumption in light distribution measurements using near field data," in *Advanced Mathematical and Computational Tools in Metrology and Testing*, 2017.
- M. Katona, I. Rotscholl, K. Trampert and C. Neumann, "Übersicht über die Berechnung von Lichtstärkeverteilungen aus Rayfiles mit ihren Vor- und Nachteilen unter Verwendung verschiedener Raumwinkeltypen," in *Proceedings of the Lux junior*, 2017.
- C. Gut, I. Rotscholl and C. Neumann, "Theoretische Leistungs- und Effizienzanalyse laserbasierter Pixellichtsysteme," in *6. VDI Tagung: Optische Technologien in der Fahrzeugtechnik*, 2014.

The publication in *LED professional Review*, vol. 63, 2017 is a reprint of the conference contribution at *LED Professional Symposium*, 2016.

SUPERVISED THESES

- M. Katona, *„Berechnung der dynamischen photometrischen Grenzentfernung"*, Master's thesis, (2015)
- S. Huber, *„Methoden zur Quantisierung kontinuerlicher Spektren für farbmetrische Simulationen"*, Bachelor's thesis, (2016)

ACKNOWLEDGEMENTS

I would like to take this opportunity to thank everyone who has supported me during the time I spend on this thesis.

In particular I wish to express my sincere gratitude to my Ph.D. adviser, Prof. Dr. Cornelius Neumann, who has not only supported me professionally, but also personally throughout my time at the Light Technology Institute and thus made this thesis possible in the first place. It was due to his excellent lecture course "Lighting Technology" that I first discovered my passion for this research area. Furthermore, I will always remember much of his advice, for instance "the three rules of management".

I would also like to thank Prof. Franz Schmidt and Prof. Wilhelm Stork for agreeing to be my co-advisers as well as Prof. Uli Lemmer for always having an open door for every Ph.D student at the LTI.

The same is true for Dr. Klaus Trampert, who has not only supported me in all scientific and formal aspects of this thesis but also generally during my time at the LTI. Many ideas and concepts originated from the fruitful discussions with him during lunch or coffee break. He also taught me the importance of clear and simple communication in science, which is - hopefully - also the case in this thesis. The constructive advice and question from Dr. Udo Krüger and Prof. Franz Schmidt have also helped to improve my work, in particular with regard to aspects of measurement uncertainties and usability.

Furthermore, I would like to thank all my colleagues at the LTI, who have participated either in discussions regarding scientific content or in diverting free time activities. In particular I would like to mention Martin Perner for his strict but always sincere and constructive feedback following oral presentations and the commitment he demonstrated during the acquisition of the near field goniophotometer and the simulation server. Further, I would like to thank Inca-Leopoldo Sayanca for constructing and providing the measurement object used for the validation in chapter 8 and, together with Tino Weiß, for their help with measurements. I am also grateful to "my" students Markus Katona and Simone Huber for many fruitful discussions as well as some humorous sketches on the whiteboard in our office. In addition I would like to thank Klaus Trampert, Simon Wendel, Martin Perner, Melanie Helmer, Markus Katona, Christian Herbold, Patric Jahn, Inca-Leopoldo Sayanca, Tino Weiß, Annie Shalom Isaac, Gregor Hollmig and all other colleagues from the LTI for the pleasant atmosphere at the institute as well as the staff and members of the Karlsruhe School of Optics and Photonics for the effort they invest in our graduate school.

I would also like to acknowledge the financial support of the German Federal Ministry of Education and Research regarding the project "UNILED 2" which enabled me to work at the LTI, as well as the professional support of "our" associated project partners TechnoTeam and OPTIS Germany.

In addition, I would like to thank Klaus Trampert, Udo Krüger, Markus Katona, Daniel Großmann, Simon Wendel and, in particular, my fiancée Julia for their time spent while proofreading the content of this thesis, and Helen Rensmann and once more my fiancée Julia for proofreading this thesis for linguistic accuracy.

I also wish to thank my parents, Peter and Beate, who have always supported me in each and every aspect of my life and studies.

Last but not least I wish to thank my fiancée Julia wholeheartedly for always sharing my joy, for always listening to my worries, for always succeeding in cheering me up or diverting me and for her endless patience with me and in particular with this thesis.

SPEKTRUM DER LICHTTECHNIK

Lichttechnisches Institut Karlsruher Institut für Technologie (KIT)

ISSN 2195-1152

Band 1 Christian Jebas
Physiologische Bewertung aktiver und passiver Lichtsysteme im Automobil. 2012
ISBN 978-3-86644-937-4

Band 2 Jan Bauer
Effiziente und optimierte Darstellungen von Informationen auf Grafikanzeigen im Fahrzeug. 2013
ISBN 978-3-86644-961-9

Band 3 Christoph Kaiser
Mikrowellenangeregte quecksilberfreie Hochdruckgasentladungslampen. 2013
ISBN 978-3-7315-0039-1

Band 4 Manfred Scholdt
Temperaturbasierte Methoden zur Bestimmung der Lebensdauer und Stabilisierung von LEDs im System. 2013
ISBN 978-3-7315-0044-5

Band 5 André Domhardt
Analytisches Design von Freiformoptiken für Punktlichtquellen. 2013
ISBN 978-3-7315-0054-4

Band 6 Franziska Herrmann
Farbmessung an LED-Systemen. 2014
ISBN 978-3-7315-0173-2

Band 7 Simon Wendel
Freiform-Optiken im Nahfeld von LEDs. 2014
ISBN 978-3-7315-0251-7

Band 8 Carmen Kettwich
Ablenkung im Straßenverkehr und deren Einfluss auf das Fahrverhalten. 2014
ISBN 978-3-7315-0288-3

Band 9 Steffen Michenfelder
Konzeption, Realisierung und Verifikation eines automobilen Forschungsscheinwerfers auf Basis von Digitalprojektoren. 2015
ISBN 978-3-7315-0301-9

Band 10 Celal Mohan Ögün
Surface wave driven molecular low pressure plasmas for general lighting. 2016
ISBN 978-3-7315-0464-1

Band 11 Theresa Bonenberger
LED Farbmischung mit chaotischen Lichtleitern. 2016
ISBN 978-3-7315-0480-1

Band 12 Michael Schöne
Diffraktive Optiken im Automobil: Achromatisierung, Athermalisierung, Formung von Scheinwerferlichtverteilungen. 2017
ISBN 978-3-7315-0613-3

Band 13 Tobias Werner
Simulation, Aufbau und Charakterisierung von autostereoskopischen Display-Systemen im Fahrzeugbereich. 2017
ISBN 978-3-7315-0617-1

Band 14 Christian Herbold
Entwicklung und Herstellung naturähnlich verzweigter Kühlkörper für LED-Systeme. 2017
ISBN 978-3-7315-0635-5

Band 15 Carsten Gut
Laserbasierte hochauflösende Pixellichtsysteme. 2018
ISBN 978-3-7315-0710-9

Band 16 Annie Shalom Samji Isaac Chandra
Intelligent Freeform Deformation for LED Illumination Optics. 2018
ISBN 978-3-7315-0741-3

SPEKTRUM DER LICHTTECHNIK ISSN 2195-1152

Band 17 Ingo Rotscholl
Spectral near field data of LED systems for optical simulations. 2018
ISBN 978-3-7315-0750-5